The Human Animal

The Human Animal

Why We Still Don't Fit into Nature

Markus Gabriel

Translated by Karl von der Luft

polity

First published in German as *Der Mensch als Tier. Warum wir trotzdem nicht in die Natur passen*
© by Ullstein Buchverlage GmbH, Berlin. Published in 2022 by Ullstein Verlag

This English edition © Polity Press, 2024

Excerpt from *The Poetry of Rilke: Bilingual Edition* by Rainer Maria Rilke, translated and edited by Edward Snow. Translation copyright © 2009 by Edward Snow. Reprinted by permission of Farrar, Straus and Giroux. All Rights Reserved.

Polity Press
65 Bridge Street
Cambridge CB2 1UR, UK

Polity Press
111 River Street
Hoboken, NJ 07030, USA

ISBN-13: 978-1-5095-5803-2

A catalogue record for this book is available from the British Library.

Library of Congress Control Number: 2024934655

Typeset in 11.5 on 14 Adobe Garamond
by Fakenham Prepress Solutions, Fakenham, Norfolk NR21 8NL
Printed and bound in Great Britain by CPI Group (UK) Ltd, Croydon

The publisher has used its best endeavours to ensure that the URLs for external websites referred to in this book are correct and active at the time of going to press. However, the publisher has no responsibility for the websites and can make no guarantee that a site will remain live or that the content is or will remain appropriate.

Every effort has been made to trace all copyright holders, but if any have been overlooked the publisher will be pleased to include any necessary credits in any subsequent reprint or edition.

For further information on Polity, visit our website:
politybooks.com

For Leona Maya
You outshine logic with life.

and the sly animals see at once
how little at home we are
in the interpreted world.*

Rainer Maria Rilke

*Excerpt from the "First Elegy," trans. Edward Snow, in *The Poetry of Rilke: Bilingual Edition* (New York: North Point Press, 2011)

Contents

Part III: Towards an Ethics of Not-Knowing   165

Introduction

We find ourselves in a complex crisis scenario. Our habitat, our natural environment as human beings, is threatening to collapse in plain view under the pressure of our modern way of life. Thanks to science and technology, we have rapidly improved our survival conditions. But, by the same means, we have made them worse even more rapidly – a dilemma that is exacerbated with every contemporary crisis.

In the meantime, the civilizational model of modernity, which consists of bringing the resource problems of the survival of our species under control through science and technology, has brought us to the brink of self-extermination. Our instruments of natural and social domination (nuclear power, automobiles, airplanes, smartphones, artificial intelligence, weapons systems, the internet, etc.) are turning against us. It is almost paradoxical. Our technological knowledge, thanks to which we have the internet, AI and social networks, is also the reason why fake news, propaganda and conspiracy ideologies are spreading like wildfire. Through automobiles, airplanes and our fossil fuel way of life, we are better connected than ever and able to interact with spatially distant cultures and people. And yet through these same means we are also destroying our shared environment and our sense of a shared reality.

It is pointless to try to cope with the complex crisis situation of late modernity in which we find ourselves by doing more of the same.[1] What we need instead is a reorientation of our image of the human being and our place in nature. That is what this book is about.

Its point of departure is a radicalization of the insight that we humans are animals. The French philosopher Corine Pelluchon has sharpened this point in a series of books with the concrete demand for a new New Enlightenment, at the center of which stands the human animal.[2]

This New Enlightenment, to which in the meantime many global thought leaders on all continents are committed,[3] starts not with reason *in general* but, rather, with *our* nature. It is essential to bring ourselves as minded living beings, i.e. as whole human beings, back into the center. We have unjustly distanced ourselves from this center in favor of a mechanistic conception of the world as a structure that can ultimately be controlled and predicted.

This, in turn, raises an old question that we have to ask ourselves again: What does it actually mean to regard humans as animals?

This question is so important because our self-image as animals makes a significant contribution to the socio-political steering mechanisms of the present and future. We can easily see this in the way we deal with pandemics and other natural disasters: disease and (human-induced) climate change are perceived as evils that can be avoided in principle, problems which should be technically remedied as quickly as possible. This has not been achieved in the case of SARS-CoV-2, let alone in the case of climate change. So far, we've handled both almost exclusively reactively rather than proactively.

Our prognostic models and approaches to solutions fail because of the challenge we face as animals that can never entirely figure out, let alone technically control, their ecological niche. We must therefore free ourselves of the illusion that our guidelines for the time of crisis and catastrophe in which we find ourselves can be obtained merely through yet another combination of science, technology, economics and politics. The daily shifting of the boundaries of knowledge does not consist in the fact that we approach omniscience. We are always learning more about what we don't know through scientific progress. (This applies to all disciplines, including the humanities and social sciences.) There is no such thing as omniscience. And there is also no meaningful way technocratically to manage the conditions of human survival in complex systems. Life cannot be contained, nor can it be predicted. The recent virus pandemic with its many variants demonstrates this impressively. We only ever know excerpts of our own form of life. The human animal cannot be overcome by technology. *Homo Deus*, which the famous historian Yuval Noah Harari envisages as the human of the future in his book of the same name, will never arrive.

In fact, this is precisely what was always known, from Socrates to the famous naturalist Carl Linnaeus, to whom we owe our species name *Homo sapiens*. Because we cannot completely comprehend ourselves, the self-models on which we depend are prone to error. Linnaeus defines the human being by its capacity to form an image of itself. The entry *Homo*, which Linnaeus assigns to the primates in his system of nature (thereby clearly locating the human being in the animal kingdom), is supplemented by the characteristic of the capacity for wisdom, *sapientia*, which, according to him, is our *summum attributum*, our most excellent quality. The human being is defined in this way by the demand to know itself. And so next to the entry "Homo" in his system stands succinctly: *nosce te ipsum*, know thyself, with which Linnaeus alludes to Socrates. The motto and the task of philosophy is and remains, according to Socrates, "Know thyself (*gnôthi sauton*)" – a saying of the Delphic oracle, which Socrates associated with wisdom (*sophia*). Linnaeus merely translates this into Latin. Because they are committed to the love of wisdom, which is a possible translation of the Greek word *philo-sophia*, philosophers are sought after wherever the question arises of who we, human beings, are.

Philosophy is about self-knowledge. This includes insight into our freedom. As minded living beings, we are free; from this follows the value of autonomy, of responsible action. At present this value is also coming under pressure in the heart of Europe, and elsewhere in the so-called free world. In order to place values such as freedom, equality and solidarity in an appropriate relationship, and thus to regain confidence in the problem-solving competence of *liberal* democracy, humans as free, minded living beings must once again be brought to the center of society. Freedom is thereby always also *social* freedom. For we are prosocial creatures, who can do nothing without doing so in association with others. Freedom and society, individual and collective, do not contradict each other. We are not freer when we are alone, for we simply cannot do most of what interests us as human beings without others. Freedom is something we realize together, not something that positions us against each other.

There is much that you and I have in common. At the very least, we share the quality of being human. Therewith we have much else in common. We have desires, hopes and anxieties, and we are embodied as finite, transient living beings. *We belong to nature.* Modern physics teaches that there are forces and natural laws which determine everything material. Insofar as we are material, embodied as animals, we are no exception to this. Modern biology and human medicine have also shown

3

us that our bodies are "animal"[4] on an elementary level and share many basic structures with other living beings.

All living things known to us consist of cells (or, like unicellular organisms, they are identical with a single cell). These in turn consist of building blocks that can be investigated biochemically and physically. These are the issues addressed by what are now known as the *life sciences* (medicine, biochemistry, molecular biology, bioinformatics, genetics, pharmacology, zoology, nutritional science, neuroscience, etc.), whose subject matter consists of processes and structures of living beings.

In the course of modernity, discoveries about the behavior of humans and other living beings have been added to physics and the life sciences, and today these are being researched in *behavioral sciences* such as psychology, cognitive science, behavioral economics and sociobiology. It turns out that we as human beings are to a certain extent decipherable and thus controllable on various levels of our existence (from the cell to social formations such as the family, friend group, or even an entire society). Many of the countless decisions we consciously and unconsciously make every day (when we eat breakfast; whom we date; for how long we wash our hands; on which side of the street we walk; whether we fall asleep on our stomach or on our back, etc.) can be scientifically explained by discerning more or less general patterns in them.

The human being is thus accessible from the **third-person standpoint,**[5] as it is called in philosophy; we are an object of natural- and social-scientific research, one object of research among others. The title of this book alludes to this dimension of being human: *The Human Animal.*

But that is not the end of the story. For in spite of the above-mentioned modern natural-, life- and behavioral-scientific discoveries about the human being as an animal, we feel that we nevertheless do not fit into nature. *The human being is not* just *an animal.* The subtitle of the book alludes to this.

We are not just natural phenomena, which one can deduce from the fact that we explain natural phenomena. The explanation of natural phenomena and so also of those aspects of our life which are irrational is, after all, not itself irrational.

This has also been pointed out recently by the famous cognitive scientist Steven Pinker, who reminds us that logic, mathematics and critical thinking are rational and were also used by our ancestors to hunt, feed themselves, and build a stable relationship over thousands of years with other groups of people and the shared environment. Human beings are and remain fundamentally rational, which does not mean that they are error-free, as the findings of modern behavioral research, psychology, and so on, demonstrate. However, to conclude from these findings that we are unfortunately not rational after all – a conclusion that would itself be rational – is not justified.[6]

Even if we have discovered in the modern age that the 'animal in us' is controlled by non-rational impulses, instincts, processes and forces, this cannot generalize to our existence in its entirety. Otherwise, this would also apply to scientific explanation itself. Thus, on the one hand, we know that our decisions are based on cognitive biases as well as on "noise," that is, on decision-making principles that are arrived at according to irrational rules. On the other hand, this knowledge, this self-knowledge, is not subject to these very cognitive biases, since otherwise we would not be able to give rational information about the limits of our rationality. This knowledge of the limits of knowledge is rather objective, backed up by scientific methods; it is arrived at from the point of view of the third person, from the outside of our experience, as it were. In short: there is objective knowledge about us as objects and subjects.

Evolutionary theory, psychology, sociology and behavioral research, and above all behavioral economics, have indeed shown how much our thinking and our individual and collective actions are determined by forces and lawful regularities that we cannot completely control. At least since the international bestsellers of the psychologist and Nobel laureate Daniel Kahneman, it has become common knowledge that we are not quite as rational and reasonable as we like to think.[7] Our impulses, desires and inner mental states are always also part of natural phenomena and are thus shaped by principles which we do not have in our hands. A part of ourselves, our 'animality,' seems simply to be controlled from the outside – by natural laws, by evolution, by society, and so on.

For example, we stubbornly persist in beliefs even when we already have information that contradicts them, which is called confirmation bias. The list of cognitive illusions or biases is long, and we all know that we have a perspective on social and natural reality that is by no means automatically correct and that we are therefore continually correcting.

Still, we can correct our limitations in collaboration with others and by working on ourselves.

The fact that we can correct cognitive biases psychologically, social-scientifically and through everyday practices of decision-making, demonstrates that cognitive biases are not a natural necessity. We are and remain free, which is not changed by the fact that, because we can only ever perceive and think selectively, we can deceive ourselves.

The question as to what or who the human being is, is by no means finally answered. For we don't know today wherein our consciousness or our rational mind consist, thanks to which we can consider ourselves as a natural phenomenon in the first place. There is not only the standpoint of the third person, the outside perspective on us, but there is also: us (i.e., for example, you and me). We share not only biochemical structures such as the human genome but also the fact that we are subjects, that is, that we each occupy our own **first-person standpoint (subjectivity)**. This includes the reality of our feelings and thoughts and also our sensory perspective and our perception of reality. The riddle of being human cannot be solved purely objectively, from the standpoint of a third person. There is no external perspective on being human that we could take in order from there to gauge the meaning of life or to see that our life has no meaning at all.

Even the most objective natural scientist, say a consultant performing open heart surgery, has her subjective perspective on what is happening. The surgeon must, after all, see the heart she is operating on and be internally composed and professional while doing so, which involves a lot of training of her subjectivity. The historians of science Lorraine Daston and Peter Galison have impressively demonstrated in their book *Objectivity* that the history of objectivity consists in designing an ideal of behavior for the scientist in order to be able in this way to recognize natural phenomena as neutrally as possible.[8] There is no objectivity without a subjectivity to match it; objectivity remains an ideal to which we aspire without ever fully achieving it in general.

This book aims at nothing less than examining the relationship between nature and mind on the basis of the human–animal interface. In the human animal, nature and mind join hands. I will call the self-investigation of human beings **anthropology**; if we consider ourselves as animals, we also speak of **anthrozoology**. Anthropology in that sense is not to be confused with the discipline we call anthropology. When in the following we deal with the human being as an animal, I am guided by confrontations with contemporary scientific findings as well as by contributions of contemporary philosophy.

The human being is the transdisciplinary theme per se. Who we are and who we want to be cannot be measured from the perspective of a single science or a single kind of science (such as the natural sciences). And the sciences that we promote as academic disciplines in teaching and research do not exhaust our humanity either. The arts, politics, common sense, the economy and the workplace, the media and our life experience are all, as human activities, also forms of human self-knowledge.

Against this background, the book addresses itself from a philosophical perspective to all who ask themselves in what being human and the meaning of life consist and how our knowledge society is compatible with the fact that we are presumably infinitely far from omniscience. Whether we like it or not, the complex crisis situation in which humanity finds itself in the twenty-first century demonstrates not only our knowledge (regarding the ecological crisis, for example) but also our ignorance and powerlessness. That is why we must rethink and adapt our actions to the new circumstances of this century, which also implies finally learning from those who were and are victims of the modern mania for control and destruction of nature. Human self-knowledge and rationality can take many forms from which we can learn, as Tyson Yunkaporta has shown in his remarkable book *Sand Talk*, in which he recommends the indigenous knowledge of the Aborigines (he himself is a member of the Apalech clan) as a dynamic model of dealing with complexity and crises.[9]

The thesis that the human being is the transdisciplinary theme per se, the theme which concerns us all, is thereby taken from the standpoint of the humanities [*Geisteswissenschaften*], which – as their German name quite rightly indicates – have the mind [*Geist*] as their subject and object. In this context, I evidently do not understand "mind" [*Geist*] to refer to a ghost or some relic of an allegedly past and vanquished metaphysical or religious thinking. **Mind** is, rather, the capacity in general to lead one's life in the light of a conception of who or what one is. That we are minded living beings is the **principal thesis of neo-existentialism**.[10]

The human being is the animal par excellence: what we know about being an animal results from our self-investigation, because we have been interested in 'animals' for millennia mainly because it is unclear how humans and animals are related to each other. Thus, in thinking about 'animals', we are always also thinking about ourselves. Thanks to our self-understanding as animals, we are the prototype of animality. The concept of an animal, I will argue, says more about humans than it does about the 'animals' from which we have been distinguishing ourselves in a false way for millennia.[11]

Because, as animals, we are a part of nature, we are interwoven with other living beings, so our actions must always be considered ecologically, in the context of other living beings and our shared habitat, planet Earth. Who we are and who we want to be also shows us, then, what we should do or refrain from doing.[12] In human self-knowledge what is and what ought to be join hands.

The currently operative view of humanity and its place in reality has reached its planetary limits. In the meantime, it is generally known and even acknowledged by economic ministers and leading economists that there are "limits to growth," as the Club of Rome report on the state of humanity and the world economy stated as early as 1972, over fifty years ago now.[13]

We must overcome the idea that scientific-technological progress as a driver of purely quantitative economic growth can be decoupled from human and moral progress. This erroneous idea leads to the self-destruction of human beings. It is an expression of a disturbed self-relationship which needs to be figured out and overcome.

We express our self-relationship individually and collectively, as individuals and as a society, in the form of strong value judgments. Visions of the good life circulate in every society. Against this background, this book deals with the **meaning of life**. Because the human being is an 'animal,' but not only an 'animal,' the meaning of life is not exhausted in the individual and sociopolitical planning of our survival. Life is more than survival.

I will try to connect the liberal pluralism of individual forms of life – which I consider sustainable even in the twenty-first century – with the question of the meaning of life. Contemporary philosophy can help with this. The American philosopher Susan Wolf has suggested that liberal pluralism ("Each one should be blessed after his own fashion") be regarded as a search for meaning *in* life. This can turn out different for everyone. However, it does not exclude the possibility that there is a meaning *of* life that we all share and on which liberal pluralism is based. The German Basic Law [*Grundgesetz*] expresses this with a strong commitment to human dignity, in which the Enlightenment tradition continues to have an impact. Reflection on our animality, on life and on its meaning has political consequences; it affects our understanding of ourselves as having dignity.

In the context of a New Enlightenment *in the age of living beings*,[14] as Corine Pelluchon recently aptly described it, we can understand the meaning of life as our moral mission. This corresponds to the currently perceptible socio-ecological awakening, which is connected with the longing to find a new commonality among human beings in order to overcome the complex crisis situation in which we find ourselves. Following Tyson Yunkaporta, we can attach the same idea of a human mission to the fact that we are "custodians of reality."[15]

The ecological transformation that humanity will inevitably have to undertake in this century will require a normative design for social change, and thus also for ethics. We can no longer rely on the false modern promise that the natural sciences and engineering, together with economists, will solve the fundamental political problems and thus relieve us of the actual normative decisions about who we are and want to be.

The self-understanding of human beings has political consequences. The palpable crisis of democracy observed by many sociologists and political scientists also stems from the fact that people demand more from politics than a clever distribution of resources. In times of crisis, what matters most is what is too uncritically called "communication." Politicians should not only communicate their decisions cleverly in order to lull the population into a sense of security, but they should also justify their decisions and in this way make the value judgments they undertake explicit. For example, the people affected by the devastating floods in the Ahr Valley in Germany in the summer of 2021 (I happen to be from that region) demand not only that political decisions be communicated well in the aftermath of the disaster but that value-driven solutions be sought together. The reconstruction of this and other destroyed regions should be done in the light of ecologically sustainable models.

To modern science also belongs the realization of the limits of scientific knowledge. We know that there is a lot we do not know. The crisis situations in which we find ourselves also consist in the fact that we have to deal with complexity, uncertainty and not-knowing. This requires an *ethics of not-knowing*.

The underlying thought that runs through this book is the idea that, through self-investigation of our animality, we can learn to recognize nature within us and outside us as something profoundly alien that we both cannot and should not master. We can never fully decipher nature and bring it under our control. We human beings depend on natural processes that we cannot concretely begin to comprehend to the extent that we could establish a technocratic paradise on Earth. We must say goodbye to vain hopes such as the idea that we could colonize other planets (such as Mars) to begin anew, or the even more far-fetched fantasy that we could upload our consciousness as software to indestructible plastic bodies, as in the TV series *Westworld*. The Covid pandemic has palpably disclosed our vulnerability and social complexity to all of us.

Of course, complexity and vulnerability existed before the pandemic. But they were, so to speak, covered over "demo-bureaucratically" (Niklas Luhmann). That is, we were not aware of them because our healthcare system in particular functioned more or less smoothly for most people. The understanding that humans are vulnerable animals, subject to ecological transformations for which they are partly responsible, must always be inherent in politics. Since we are not only animals but also specifically minded living beings who possess ethical insight into moral contexts, anthropology, ethics and politics are inextricably intertwined in the human being.

To be sure, the New Enlightenment calls for us to use more scientific disciplines (above all including the humanities and social sciences on an equal footing with the natural-, life- and behavioral sciences) in order to develop an adequate self-portrait of the human being. Beyond this, however, it also calls for an ethics of not-knowing, based on the recognition of the fact that we are far from living in an age of complete knowledge and mastery of nature. The unquestionably impressive and in part desirable advances in knowledge and technology in the modern era must no longer obscure the fact that there is an indeterminate, perhaps even infinite, amount of facts that we do not know and will never know. However great our knowledge of nature is, our not-knowing is even greater (which we even know today, since it has become clear that the observable universe consists of 95 percent dark matter and dark energy, which we cannot directly explore experimentally). Reality exceeds our knowledge claims. This is not mere conjecture, but something we know. So we do really know that there is a lot that we do not know.

As we shall see, Socrates was right, though he did not say that we know nothing but, rather, that we can become aware that we do not know many things. Socrates' wisdom is a form of knowledge, not a skeptical cult of ignorance. The human being as an animal can become aware of its own not-knowing. This is what Socrates called wisdom. And in this Carl Linnaeus followed him with his famous definition of the human being as *Homo sapiens*. For the Latin sapiens means "capable of wisdom," not "wise." The human being, as a minded living being, is the philosophical animal that can historically change both itself and non-human reality by forming an image of itself. So let us set out together to explore our humanity and animality, the meaning of life, and the depths of our not-knowing!

PART I

WE AND THE OTHER ANIMALS

… the same incomprehensibility, as to how connection is possible between matter and mind, continues to oppress us ….
We leave behind man, as evidently the most devious problem of all philosophy …

Friedrich Wilhelm Joseph Schelling

Since time immemorial, human beings have been concerned with the question of who or what they actually are. In order to give an answer to this question, we have to distinguish ourselves from something else. For to define something is to distinguish it from something other. So one of the oldest known definitions of human beings is that we are animals gifted with reason and language. The human being, according to Aristotle, is the *zôon logon echon*, the creature that has *logos* (language, reason), which in Latin became the *animal rationale*, the rational animal.[2]

Before we take a closer look at this feature of language and rationality that is supposed to distinguish the human being, it is important to note that the oldest known self-definition of humanity, thus the oldest **anthropology**, already determines the human being as a living being or as an animal, as *zôon* or as *animal*. Already before Aristotle, in the phases of humanity's mythological consciousness, the human being is classified by itself as somehow part of the animal kingdom.

The biblical picture of humanity is an exception to this, as it strongly separates humans and animals from each other. Only the human being, as the image of God, is an outstanding inhabitant of paradise. Admittedly he is regarded as the inhabitant of a garden, as a part of nature, yet not only as the crown of creation – just think of the story of Noah's ark – but also as the being who is responsible for the animals.

As a first observation, we can state that humanity has been conceived as a living being or as an animal long before the insights of evolutionary theory and the rapidly progressing differentiation of the life sciences (medicine, biochemistry, molecular biology, biocomputing, genetics, pharmacology, zoology, dietetics, neuroscience, etc.) that followed. Indeed, it is the default situation in human intellectual history for humanity to define itself as an animal.

Over millennia, the self-image of humanity has developed mainly because we have changed our image of animals. Through the progress of natural sciences and technology in modernity we have recognized how complex natural history is. In the nineteenth century, especially in the wake of the Darwinian revolution, we gained the insight that the manifold forms of life and thus also animal species are subject to processes of natural selection. Insofar as we are animals, we adapt to our environment on the different levels of our life, which we of course always shape at the same time. Nature, to which we belong, is not a static container but something that we change through our life processes. And today we know that this nature to which we are adapted has been modified for billions of years by life processes of other life forms, e.g. by the cyanobacteria that have been producing oxygen in the atmosphere over millions of years.

In the nineteenth century, alongside the pioneering theory of evolution, the view also prevailed that we understand our animality as something that is alien to us and that is beyond our conscious control. The then newly emerging sciences of biology, psychology and psychiatry held up a mirror to us, in which we didn't see just mere rational actors (who react and act with consistent rationality) but also animals.

The impressive findings in the life sciences over the last two centuries have led us to believe that we are puppets of our life processes. So it is common today to believe that our genes, our instincts, our neuron circuits, or our biologically determined personality controls our decisions. For example, the renowned neuroscientist Eric Kandel, who was awarded the Nobel Prize, believes that what Freud called the unconscious can now be understood as an evolutionary inheritance using life-science methods.[3] According to this view, it is not just that we are not "masters of our own house,"[4] of the life of our own *psyche*, as Freud speculated, but we are even strangers in our own body, which produces our conscious experience, our ego, at best as a control center, in order in the ideal case to adapt even better to the environment.

We should note that Freud only apparently dethroned the human being. He only identifies the famous Id with the 'animal in us,' as it were, and beyond that attributes a complete psychic apparatus to us human beings. Although, like Darwin, Freud wants to include us in the "animal series,"[5] he thinks that the other animals hardly have what he calls the Ego, not to mention the Superego. Freud's conception of our psychic apparatus follows the model of demarcating the animal in us (the

Id) from our rationality (the Ego and the Superego), which in his eyes, of course, is constantly disturbed by our being animals, by our drives, which is why he famously conjectured that our civilization is characterized by discontent.[6]

The history of our picture of nature, the body, animals and humans, which I can only touch on here, shows that, on the one hand, it is normal for humans to define themselves as animals but that, on the other, the **image of humanity** also varies with the respective formulation of the concept of an animal. The interdisciplinary field of research, recently characterized as *Animal Studies* and **Anthrozoology**, deals with exactly this topic. Depending on what we represent as an "animal," our image of ourselves, i.e. our image of humanity, changes, and this in turn has ethical consequences. If, for example, one rightly insists that we need an animal ethics but then essentially thinks only of mammals and ignores insects because the concept of an animal is not understood sufficiently broadly, one will act quite differently than if, conversely, one considers almost all self-moving, non-plant forms of life (perhaps even bacteria) already to be animals with which one must deal ethically.

The sentence "The human being is an animal" accordingly needs more exact consideration – and that's what we're about in this part of the book. For it is decisive in the first place what we understand by "living being" or "animal" when we classify ourselves in this way as a natural phenomenon. And, moreover, it is at least as important to think more exactly about the further characteristic by which we differ from the other living beings or animals with which, as we say, we share our animal being or animality.

The Logical Animal – How Humans Became Animals

The history of logic is decisively pushed ahead by the self-definition of the human being as *animal rationale*. As mentioned, Aristotle defines the human being as the living being that has *logos*, from which the idea of logic is derived.

Logic is the basic discipline of philosophy, whose object is rational thinking itself. It is concerned with the question of how we think correctly (i.e. rationally), i.e. how we ought to think, and its form of implementation is thinking about thinking. The point of Aristotle's definition of the human being as an *animal with logos* is that this also makes us a *logical animal*: to have a *logos* means to be able explicitly to distinguish oneself from other things, i.e. to have knowledge about oneself and other beings.

Since Aristotle, logic teaches that the textbook case of a definition consists precisely in the fact that one first has to define a genus (Greek *genos*, Latin *genus*) to which something belongs in order to distinguish it further from other species (Greek *eidê*, Latin *species*). The origin of the scientific division of living things into genera and species is the logic of the ancient Greeks.

Since the ancient Greeks, the phylogenetic tree of life has been thought to be grounded in a supreme genus represented by divine living beings, which the human being is supposed to imitate in order to become like them. In this tradition, the order of genus and species is considered a *hierarchy*: the higher something is in the hierarchy, the closer it is said to be to the top of the value order, to the perfect being or to the good. Incidentally, hierarchy literally means "sacred order," and the idea comes especially from angelology, i.e. from the late antique and then medieval discussion of how the human being relates to angels and ultimately to God as the source of all being and thought. The classic in this field is the writing *De coelesti hierarchia* by the so-called Pseudo-Dionysius of Areopagita. It dates from the fifth or sixth century and investigates the order of the angels. The author of this writing, whose true identity is unknown, coined the concept of hierarchy, which thus always exhibits a theological dimension. For thousands of years, humanity has defined itself not only in contrast to the animal (or to the other animals) but also in contrast to divine beings.

Even today, this aspect should not be brushed aside with the naïve reference to a widespread secular view of humanity, which regards us as a more or less normal element in the series of natural phenomena. For, even in our times, many if not most people believe that we occupy a privileged position in the animal kingdom and even that we have an immortal soul or that we will be reborn, that we therefore are in no way identical with our animal bodies, which perhaps will one day be able to be described completely in the language of today's life sciences.

The concept 'animal,' which divides the realm of living things into genera and species, is therefore by no means value-neutral, as illustrated by its historical origin, its genealogy. Rather, from the outset it contains valuations which obtain on different levels. For example, it distinguishes between normal exemplars of a species and their deviations, so that the concepts of human being, lion, pigeon, etc., are understood as normal forms from which deficient exemplars deviate – a view that is rightly criticized by theories of gender or within more recent *disability studies*.

The Greek philosophers basically assumed that the fundamental principles of the order of reality are in their essence more powerful and also better than that which depends on these principles. Thus, for example, elements such as water, fire and air were placed at the top of the order of being as fundamental, while the individual things found in nature were conceived as manifestations of a deeper structure of fundamental elements that is more powerful than we are. This still reverberates today in the thought that the laws of nature, like divine laws, dictate what can happen. In the end, nothing and nobody could escape the force of gravity. The Greek concept of a principle (*archē*) means not only beginning but also empire. Life-scientific concepts of order and ideas of power are traditionally narrowly connected. This is still deeply inscribed in our thinking up to today, e.g. when we understand ourselves as the product of evolution and think that we are at the top of development, at any rate in a negative sense, as we hold that nature as a whole depends on how we live.

---

*What distinguishes a species from all other species within the same genus is generally called the **specific difference**. Here, we are interested in a specific specific difference (a typical philosophical nesting, not a printing error): the **anthropological difference**. It consists in how and in what humans differ from other living beings.*[7]

The point of departure in this part of the book is the idea that the human being turned itself into an animal as soon as it started to distinguish itself from other living beings in a specific way, i.e. by defining itself. The classical concept of defining something is to identify a specific difference of the thing in question that sets it apart from similar things in the same genus. In this way, a definition carves out a species from a genus. As you can see at a glance from this vocabulary, the very definition of a definition since antiquity has involved the concepts of genus and species, what Plato and Aristotle, the inventors of the concept of a definition called the *genos* and the *eidos* respectively.

By defining itself, the human being introduces the idea that it is an animal plus something else (say, for example: language, reason, mind, immortal soul). To be human thus means to be a particular animal, an animal that is special in virtue of a specific something. The human being *is* an animal in light of the fact that it conceives itself as one. To paraphrase a dictum of the great existentialist philosopher Simone de Beauvoir, one might say that the human being is not born but, rather, becomes animal.[8]

To be sure, we do not first have to become living beings but are already human in some sense during our development in the womb, because we have a typical human DNA, which enables developmental steps that are first carried out in the womb and thereafter in larger communities. This expresses our biologically describable form of life. Whether this can be determined on the basis of a cell structure or a time scale that determines from when exactly a fertilized egg cell becomes a human being is a difficult and ethically significant question that I do not wish to pursue here. The only thing that is certain is that we are human beings many months before we are born, so that being human is not something that can first be learned and which we can therefore also fail at. Our humanity is inalienable and cannot be lost. On the other hand, our being an animal, i.e. our conception of ourselves as animals, is historically contingent. That means other conceptions of the human being were and are possible.[9] For each conception of the human being as animal depends both on a specific concept of animal and on a specific concept of the specific something that accounts for the difference between us and the other animals.

The Specific Something

The human being is not born as an animal, nor does it develop into an animal in the womb in order to become part of a culturally and mentally formed human society through social training. Rather, as soon as a human being is a living being at all (again: starting in the womb), it is already a human being in the full sense. Thus, a human being doesn't come into the world as biological hardware on which a cultural software has to be installed through education. Accordingly, human dignity is bound solely to being a human being and not to any other condition – such as the development of special abilities and higher capabilities. Everybody who comes into being by way of the usual evolutionary channel of sexual reproduction for living beings like us is a human (and has therefore grown in a mother's womb for some time; at least, this is still state of the art).

In this respect, one can already see a shortcoming in Aristotle's classical definition of the human being as distinguished from the other animals by the features of reason and political organization (which, along with reason, he considered characteristic).[10] For many human beings (certainly those in the womb) are in his sense neither rational nor political animals. Aristotle's definition of the human being therefore leaves out too many of us. Nor does any other traditional view of the human as animal cover humanity. No attempt to hit on the essence of the human being by way of identifying an outstanding feature, a *definiens* of humanity that clearly distinguishes us from all other animals, has ever been successful.

The human being is not in itself an animal but becomes one through its self-definition. In the context of answering the question who or what we are, the human being separates off a part of itself – animality. This separation is the origin of the erroneous concept of an animal kingdom, from which the human being stands out by special characteristics, by its specific difference.

By defining itself, the human being distinguishes itself from other animals. It understands itself as *animal plus something*. According to this logic, the specific something that distinguishes us must be missing in the other animals, because otherwise we would not have properly distinguished ourselves from them. This then means that other animals are humans minus something. They lack what we have got (even though they have their own specific something).

21

The real difficulty here is not, as many today believe, that there is in fact no special characteristic of being human (on the contrary, there are quite a few such characteristics) but, rather, that we identify in ourselves a core of being animal, the concept of animality. We share this animality with the other animals and yet we are supposed to distinguish ourselves from them by that specific something. But what exactly is the animality we are supposed to share with the other animals if our concept of an animal is that of a human minus our specific something? How can we share such a deficiency with other animals?

If the specific something consists in language and rationality, which the other animals lack *per definitionem*, then our own animality must also lack these capacities. The 'animal in us' therefore appears as an irrational part of the soul, a notion explicitly advocated by Plato, Aristotle's teacher, and prominent until today's idea that our bodies have evolutionarily old, irrational, purely animal parts (such as the amygdala, a part of our brain associated with certain instincts of flight when confronted with physical danger).

Animality is defined by the self-definition of the human being on its own part. For we delimit it in the self-definition from our specific something. But this means that the animality by which we define ourselves in this way is already deficient, i.e. it represents a lack, as it is not the whole human being. In short, the concept 'animal' that we obtain through the self-definition of the human being as animal plus something generates the idea of an animal kingdom to which all animals belong in the same way, but which are also distinguished from it by specific characteristics. Thus we project a part of our self-definition, our deficient animality, onto the other living beings. In contrast to the human animal, the animal-animals are defined, so to speak, as deficient entities, which lack that which distinguishes us. This is a consequence of our traditional self-conception as animals and not, as some believe, an artifact of a modern conception of rationality.

In a second step the concept of our own animality already contains an implicit or explicit valuation. For that which distinguishes and defines the human being has been regarded for millennia not only as a quality we have but as a special source of value. And quite rightly, for it is true that we humans are qualified for a special kind of moral reflection, i.e. for ethics.

*In general, **ethics** can be understood as a sub-discipline of philosophy that deals with the question of what we should do or refrain from doing insofar as we are human beings.*[11]

Because we are capable of insight into the foundations of our actions and are able to define what we are like as human beings, we are also able to change. How we define ourselves as human beings helps determine who we are. **Philosophical anthropology** (any attempt to answer the question concerning our essence or nature as human animals) is a crucial source of value knowledge. We are human beings even without a definition, so our human dignity does not hinge on any definition or image of humanity. However, there are images of humanity that cast doubt on our human dignity or dehumanize people, which is why it is important critically to observe how exactly people distinguish themselves from other things or living beings. Our self-conception as animals reveals elements of our value system.

The American psychologist Barry Schwartz puts it in a nutshell when he states in his booklet *Why We Work*:

Ideas or theories about human nature have a unique place in the sciences. We don't have to worry that the cosmos will be changed by our theories about the cosmos. The planets really don't care what we think or how we theorize about them. But we do have to worry that human nature will be changed by our theories of human nature. ... [I]t is human nature to have a human nature that is very much the product of the society that surrounds us.[12]

So it is absolutely true that we humans have a special responsibility for ourselves and other living beings. While lions do not consider becoming vegetarians, and chimpanzees do not work to establish gender justice by institutionally implementing the legitimate concerns of feminism, we humans are capable of such fundamental revisions of our social systems and, so too, of moral progress. We are therefore not only the problem that fuels climate change and environmental degradation, as many climate activists and anthropological pessimists believe, but the only solution that can save us. The other living beings will not work on their CO_2 emissions or change their consumption habits. It is in our hands, and it is our responsibility to revise our view of nature, animals and human beings if we are to succeed in working together to ensure that as many people as possible on this planet can lead a life worthy of human beings in this highly industrialized, globally networked, technological modern age. Despite all modern progress, we are still far away from this goal, whereby we must always keep in mind that modern advances, which freed millions from extreme poverty, have also brought about climate change, environmental destruction and weapons of mass destruction, and thus new forms of poverty in the modern age in the first place. Modern famines are also the results of modern scientific and technological progress and global supply chains, which are closely intertwined with neoliberal ideas that global trade somehow provides for humane progress by itself, which is simply wrong, as recently proven once again by Russia's brutal war of aggression against Ukraine. Global trade, unfortunately, does not automatically lead to eternal peace.

Our ability to be ethical does not raise us above the other living beings. It does not establish a claim to leadership or power on planet Earth but is at most the basis for us to deal responsibly with one another and with other living beings who share our habitat, i.e. the surface of planet Earth, with us.

The anthropological difference, by means of which we distinguish ourselves from the 'other animals,' is itself not biologically given but is set by us. The ability to determine such differences, thus to have a self-image, is given to us humans. This circumstance implies that we are the source of knowledge for objective values, which other living beings do not grasp, but from this it follows that we can recognize what we owe to other living beings, i.e. how we should behave towards them. Because we can do otherwise (we can change our self-conception and thereby create new abilities), we should behave in such a way that we protect other living beings. Animal ethics follows automatically from the fact that we are capable of moral insight, which also means that we have to ask ourselves whether, for instance, we are allowed to kill mosquitoes at will (because they attack and endanger us, for example)

24

or, as in Denmark in the fall of 2020, to kill millions of minks because they are infected with the coronavirus and we are afraid of new variants (which came from humans and not from the minks!). But to our pets or genus-related primates we should please be very kind … Our everyday animal ethics is largely anthropocentric because we sympathize with some 'animals' and not others. This shows the projection mechanism that I criticize in the following. Animal ethics can only make systematic progress when we overcome the prevailing concept of animals or transform it to overcome the anthropocentrism that manifests itself in our sympathizing with 'animals' with whom we coexist on the same scale (the mesoscopic scale on which we move), while at the same time we as humanity are responsible for the greatest extinction of species in many millions of years.

Nature is Not a Safari

In the context of its self-definition, humanity easily overlooks the fact that the other animals, in contrast to human beings, have by no means enough in common so as to be essentially all the same in their animality. We are not in a kind of safari organized by nature, in which we ask ourselves from a more or less safe observer position how exactly we belong to the animal kingdom while simultaneously rising above it.

In the wake of the Darwinian revolution, which made it possible to explain the origin of animal species according to principles from which the life sciences later emerged (mainly thanks to microbiological breakthroughs), there has been surprisingly little change in this state of humanity's self-definition. Whoever today expresses the thesis that the human being is an animal usually understands this to mean something like the assertion that we are descended from apes (or, of course, more precisely: that there are evolutionarily related ancestors of human and non-human primates). According to this view, we are on some branch of evolution, which all in all leads back to archetypes of living beings from which all known animal species have evolved by variation of genotype and phenotype, and which can be investigated today with great precision. But this view of the human being as an animal is not a scientifically well-documented insight into our nature. For the evolutionary biological research of animality does not explain the human being any more than the geological research of Mars explains Michelangelo's *David*. What it does successfully explain are the causal details of the interplay of various cellular systems in those organisms that we identify as exemplars of humanity. Without life processes at the cellular level, there would be no human beings. However, it does not follow that we are identical with these processes, that we are thus complex clusters of cells. The life sciences only get human beings into view if we have a prior conception of ourselves. We cannot derive such a conception from research into our cellular organization alone.

Our animality, our being an animal, is based on biological processes thanks to which we live and survive. These biochemical, physical processes are also found outside of the human being. But because we always have to define ourselves in order to be animals, we do not merge into these processes. To be an animal and to be alive in a scientifically recognizable way are two different things. Of course, this also means that images of human beings that seek to detach our humanity from these processes and our animality (e.g. by identifying us with an immortal soul that lost its way in a human body) are false.

The question arises as to how our image of ourselves as animals relates to our human biological knowledge, which is by no means closed and conclusive, as the biological systems that make us human animals are not fully knowable. For example, the human nervous system (paradigmatically, the so-called brain, which is closely interwoven with our mental processes) is in itself far too complex and dynamic for us ever seriously to hope to see through its workings in such detail that we could predict or even control its dynamic development.[13]

The thesis that the human being *becomes* an animal by its self-definition as an animal does not mean, of course, that the theory of evolution, molecular biology or human medical research is in error but, rather, that the elementary processes of the living, which are investigated by the life sciences, do not exhaustively capture our humanity. The human being (like other living beings) is always more than a particularly complexly organized heap of cells. A view of being human and animal which reduces us to cellular processes results only from a misguided abstraction from the fact that we are *minded* living beings, beings who don't admit of being fully identified with their biologically measurable organic constitution. Our human mindedness is far from fully deciphered. As a matter of fact, we do not know in what exact way human mindedness (which involves objectively shared, cultural meanings, traditions, languages, etc.) is related to the evolution of our bodily features.

Aristotle thought that we can determine the nature of human animality by the fact that "human begets human."[14] While human beings and lions cannot interbreed because they cannot produce common biological descendants through sexual reproduction, this does work between humans. For Aristotle, the ability to reproduce within a species constitutes animality, by which we are not distinguished from other animals. For even lions and giraffes are known (at least so far) to be incapable of interbreeding. This view is still spread today as a standard for the definition of species ("species = reproductive, genetically isolated units").[15]

But at this point Aristotle gets into real trouble, and with him large parts of contemporary human sciences (which have naturally advanced greatly in evolutionary-biological terms since antiquity). If one goes through the human exemplars, it is not just children who are incapable of procreation but also many adult humans. It is therefore wrong to believe that our animality, our belonging to a species, can be defined by the fact that we humans are capable of procreation, especially since this also applies to other living beings and does not distinguish humans. Young human beings are not capable of procreation and many other young living beings are not either. In addition, it holds for living beings that not every individual is procreative or capable of procreation. Thus, the idea of tying our biological identity to the capacity of procreation does not apply to all of us and, thus, is wrong.

Aristotle's view that we should define species within the framework of a theory of self-reproduction, i.e. the propagation of similar forms, still has an impact today. *Genealogy*, which deals with the general origin of forms, later became *genetics*, which is impressively well documented in natural science and thanks to which we know the details of the blueprints of organisms.

Mind you, Darwin could not concretely solve the **problem of speciation**, i.e. the formation of species, because he "knew nothing about genetics," as established by the American philosopher Daniel C. Dennett.[16] Dennett notes that the criterion according to which we are dealing with two species in the case of two life forms if they cannot be interbred (i.e. if they cannot produce common descendants through sexual reproduction) is "vague at the edges."[17] In any case, the **criterion of reproductive isolation** is not sufficient to solve the problem of speciation. Even so, since the connection of Gregor Mendel's theory of heredity with modern genetics (which finally proved the existence of genes) it is considered as an essential building block of **neo-Darwinism** (i.e. Darwinism plus genetics). Neo-Darwinism is new in that it combines genetically explorable lines of inheritance with Darwin's idea of natural selection, which is now mainstream in the life sciences.[18] Moreover, it is controversial because it is often paired with the outdated notion of the selfish gene, i.e. the idea that we are actually lone wolves who want to pass on their genes to the next generation. In contrast, the physician and neuroscientist Joachim Bauer recently argued that we could possibly already talk about empathy at the level of genes.[19] Genes engage in cooperative problem-solving and are not selfish alone.

While genetics as well as what are now called life sciences deliver completely accurate research results and thus factual findings (facts), they do not yet solve the riddle of how we can meaningfully define ourselves as animals. For the only seemingly *neutral idea* – the idea that animals are those self-organizing and reproducing living beings that are divided into species that can produce offspring with each other – implicitly or even explicitly formulates a *normative ideal*. It says that our animality is measured by our procreative capacity. Whoever cannot procreate would therefore not be an animal and not a human being or, in individual cases, an animal and a human being with a deficit.

At this point, a central argument from modern *gender theories* comes into play: the self-conception of the human being as a procreative animal of a certain species directly or indirectly expresses a normative ideal of the organization of human society.[20] From the conception of the human being as a procreative animal, very effective forms of ideology and a social reproductive order have emerged which normatively prescribes how we humans as animals are at all successful. Up to today, the definition of the human being as an animal (in turn defined by means of reproductive abilities) leads to ideologies and contributes to ludicrous ideas of human sexuality – one thinks, for example, of fundamentalist-religious discussions about contraception.

---

*The fact that contraception can in principle appear as a moral problem only results from the fact that reproductive ability and willingness is used to define the human animal, which is a case of the **projection structure** described below. First, the human being establishes its own animality and looks for characteristics it has in common with the other animals. This then gives rise to the concept of the animal in us (the animal in general or the brute). In the next act, being an animal is then applied again to humanity itself, either to domesticate it through culture and civilization or normatively to identify with it (by rejecting contraceptives, for example, and elevating the reproductive capacity of our animal species to an ought).*

The Anthropocene as Hubris

The modern development of biology as a life science has led to the fact that today, in the light of certain research results, the human being is classified in the animal kingdom. Since Darwin at the latest, many people consider it a foregone conclusion that we are animals in the same sense as all the other living beings that we classify as such.[21] Humans are one species among others. Just like all other forms of life, they have evolved according to the same principles that can be investigated by natural science, especially molecular biology. This is correct.

Nevertheless, humans differ from all other animals, and not just zoologically or genetically, i.e. not only by the fact that humans can only produce offspring with other humans (which does not hold for all individual humans anyway, as noted). Human beings do many things that no other creature known to us is capable of: we travel in airplanes to take beach vacations; cultivate wine; pursue sciences and the humanities; work in open-plan offices; exploit people to satisfy our consumer greed; write novels; defend feminism and the rights of transgender persons; hold general elections; program computers; learn foreign languages; and read this book.

29

The human being is different from the other living beings in an almost infinite number of ways. These living beings are in turn just as different among themselves. Bacteria, mushrooms, giraffes, dolphins, bats, trees, etc., can by no means simply be classified into a homogeneous plant and animal world (an ontological flora and fauna). Bacteria and bats do just as different things, have just as different abilities, as human beings and meadows. Bacteria can exist in hot places in the deep sea and produce oxygen; bats perceive reality completely differently than we do. Our intelligence and problem-solving abilities are adapted to our environment, from which we are not allowed to infer that they represent some absolute norm. Lichens, which are quite widespread and long-lived, do not fit into the 'plant, animal, human' scheme any more than bacteria do – not to mention viruses, which play an essential role in the realm of the living without seeming quite to belong to it.

Therefore, we should question the idea that any living being is *only* an animal, i.e. an instinct-driven natural occurrence that is exclusively interested in its own survival or the survival of its fellow species members. The fact that the human being is by no means *just* an animal is accordingly just as much a finding of modern evolutionary biology as the fact that our organism develops according to molecular genetic blueprints that are quite similar to those of other living beings.

The debate about how the human being differs from other animals, even though we belong to the animal kingdom, is by no means over. In fact, it was particularly important to none other than Charles Darwin himself to elaborate the difference between humanity and the other animals without compromising his newly gained insight into the origin of species. In the fourth chapter of his work *The Descent of Man* he treats in detail "the mental powers of man and the lower animals" and opens his considerations as follows:

> I fully subscribe to the judgment of those writers who maintain that of all the differences between man and the lower animals, the moral sense or conscience is by far the most important. This sense … is summed up in that short but imperious word ought, so full of high significance. It is the most noble of all the attributes of man, leading him without a moment's hesitation to risk his life for that of a fellow-creature; or after due deliberation, impelled simply by the deep feeling of right or duty, to sacrifice it in some great cause.[22]

Immediately following this passage, Darwin continues with a quote from Kant's *Critique of Practical Reason*.[23] Thus, somewhat surprisingly, he explicitly positions himself in the Kantian tradition, which conceives of the human being as a moral being receptive to the categorical imperative. The categorical imperative states that we should always act in such a way that our action could become an absolute rule, a law that applies to everyone. The various formulations Kant found for this imperative boil down to the basic notion that morally good actions must be universally valid, i.e. independent of the individual interests of a single agent. Morality thus goes far beyond altruism, because one can still be self-interested in altruism for the other's sake, whereas morality for Kant demands that we are able do what is objectively right completely without self-interest.

Darwin is now saying that there are also preforms of moral normativity and thus of an unconditional ought in other living beings, the animals, but he clearly distinguishes human beings from them. He claims, for example, that dogs possess "something very similar to conscience,"[24] and that, in general, "the so-called moral sense is aboriginally derived from the social instincts."[25] On another note, Darwin also shares the racist judgments of his time against the "savages," which run through the entire work. He counts the "savages" as human beings only to a limited extent, and in a certain way they come off even worse than the animals, but that is another matter.

If even Darwin, the founder of the theory of evolution, is of the opinion that the human being does not really fit into the animal kingdom, then this shows at least that the relationship of our picture of humanity to our picture of animals cannot be regarded as clarified simply by referring to Darwin and the theory of evolution. For Darwin by no means placed us unambiguously in the animal kingdom, as he is often represented today.

As the complexity researcher Dirk Brockmann stresses, the fact that Darwin's concept of animals and plants is based on observations that concern "only a small section of the world"[26] is an additional complication for any account according to which Darwin is credited with the discovery of our true animal nature:

His chains of reasoning refer to phenomena observed in "large" animals and plants. The entire microbiological world remained hidden from Darwin.

And if we remember that the diversity of species among microorganisms (bacteria and archaea) is about 100,000 times greater than among all plants and animals, the theory is based on a marginal group of life forms.[27]

That the human being is an animal seems to be completely self-evident nowadays; it is common knowledge. On the other hand, this apparent self-evidence is partly implicitly, partly explicitly questioned by life science research, which dissolves the concept 'animal' on closer examination.

More likely implicit or concealed, this common knowledge is obscured by the fact that humans are imagined as a particularly important factor of planetary conditions or as the top of a family tree. In the age of the climate crisis, for example, many circles are talking about the "Anthropocene," which goes back to the chemist and Nobel Prize winner Paul Crutzen (who died in 2021). According to this notion, in the **Anthropocene** era, humans have become the decisive geological factor, which is measurable, among other things, through the impact of human-made climate change that has long been felt by most people.

Due to human activities, species have been dying out for millennia, biodiversity is thus decreasing, and processes on many levels of the evolution of living things are thus changing. Of course, it is true that we are responsible for the massive damage to our own livelihoods, because we fuel human-made climate change day in and day out with our way of life, our traffic, our industry and unfortunately also with our terrible wars. Often we do not even notice this. If you watch a documentary about climate change on a streaming service, for example, you emit about as much CO_2 in half an hour as someone who drives 6 to 7 kilometers in their gasoline-powered car (according to at least some calculations).[28] Click by click, like by like, search query by search query, we contribute to global warming, not to mention the gigantic amount of energy we need for the servers thanks to which humanity is now globally networked on the internet.

It is thus a bitter irony that the supposedly most highly evolved species is on the path of its own extinction and even knows this. But in this connection the point of the theory of evolution is that no species forms the top of a family tree, for evolution is not a line but a complex fabric of different developments which we in no way control. There is no top of the hierarchy of species, as there is no such hierarchy in the first place! It is therefore already human hubris to believe that we live in the Anthropocene. Humans are not important for the planet or for life on it. The planet itself does not notice if we live in the Anthropocene, and Nature (so-called) does not care what kind of biodiversity there is. Strictly speaking, formulations such as "nature is indifferent to us" or "nature is cruel because it is not interested in us" are already incorrectly anthropocentric, as nature is neither indifferent nor not indifferent to anything. Human-made climate change is a problem for human beings and for many other living beings, but it is neither the destruction of nature nor the beginning of the end of life. We cannot destroy nature, and we are neither the beginning nor the end of the story of life.

For good reason, the term Anthropocene has not yet found its way into the official knowledge of the earth sciences. It is a popular but by no means fully recognized as a geological term. The postmodern French sociologist of knowledge Bruno Latour sees this as an omission in which a backward conception of the Earth finds expression.[29] Bound up with the expression "Anthropocene" is an image of the human being's key geological position, which still thinks along the lines of humanity's exceptional position. Insofar as we ascribe to ourselves the ability to pose a threat to the planet as a whole and all living beings on it, we overestimate ourselves beyond proportion and paint a picture of the biosphere in which we are still the – albeit diabolical – crown of creation. Thus the well-documented phenomenon of human-fueled global warming should not be misused as the basis for a perverse narrative. The human being is neither the crown of creation nor its destroyer. Of course, this is not to deny the fact that we are massively endangering or completely destroying our own livelihoods, and those of many other living beings, nor that we are entirely responsible for the mass extinction of countless species.

Equally widespread, yet misguided, is the idea that there is ultimately a single evolutionary chain of species, at the end of which stands the human being. There is not one single line of evolution but many branches and variations, all of which do not seem to follow an overarching principle. From a biological point of view, the evolutionary paths of the many life forms presently discoverable can at best be measured by how successful, i.e. how enduring, they are. If they are classified according to such a criterion, the human being is certainly not a particularly successful living being. Many of the living beings that exist today, such as bacteria, fungi and plants, will presumably still exist long after the extinction of humanity. So, if something or someone could be considered as the crown of creation or the pinnacle of evolution, it would be microbiological life forms rather than any of those creatures that we commonly classify as animals.

In any case, it is certainly not yet clear whether, from a biological point of view, the epoch in which there are human beings is even remotely as significant as it appears to us. Nevertheless, as we will see, the human being stands in the center, namely in the center of minded or spiritual life. Our capacities of minded self-determination and thus of self-definition are exceptional phenomena indeed. Therefore, we should also pose anew the question of the meaning of life and not attempt to answer it exclusively or primarily in terms of the life sciences. This is what the second part of this book will be about.

The Network: Plants, Bats, Fungi

In order to get rid of the misconception, on the one hand, that the human being is an animal in the biological sense of the word and, on the other hand, that we have particular (positive or negative) importance for our planet, we can set up a simple thought experiment. Let's ask ourselves with which life form an extraterrestrial intelligent civilization observing our planet would come into contact. Provided that the extraterrestrials are not accidentally shaped like us humans (as naïvely imagined by most science-fiction fantasies), they might, upon closer analysis of the life forms they find with us, come to the thought that they should communicate with plants rather than animals. For plants make up an impressive biomass and, on many levels of which we are not usually conscious, they are the actual shapers of the processes of the living – a theme that is playing an increasingly important role in contemporary philosophical and biological contemplation.[30] Perhaps the aliens would also be more likely to communicate with bacteria or fungi, because these are found everywhere, even in the deep sea or in the stratosphere, and make an essential contribution to the natural history of the living.

If we put ourselves in the position of non-animal life, the human hubris that we are the pinnacle of evolution in a positive or negative sense easily evaporates. The various forms of life that we find on our planet form networks in which there are no biological hierarchies. The reality of life consists in the fact that it is teeming with forms, which led the philosopher Gilles Deleuze and the psychoanalyst Félix Guattari as early as the 1970s to start from *rhizomes*, i.e. subterranean networks, instead of family trees in order to describe the developments of life.[31] Life proliferates and produces interconnections that can be as complex as the human nervous system, which appears increasingly complex the more closely neuroscience puts it under the microscope.

Living beings form a network – an insight, by the way, at least as old in philosophy as Plato's famous doctrine of ideas. In his dialogue *The Sophist*, in which Plato investigates the question of how appearance, error and non-being are actually conceivable and can be overcome by philosophy, he puts forward the thesis that our thought-forms – for which he momentously coined the word "idea (ἰδέα, εἶδος)" – form a living network, an internet or *symplokê* (συμπλοκή). This not only makes Plato's theory of ideas a precursor of the internet but, in particular, indicates that he recognized life (ζωή) as a wild growth from which forms spring.[32]

In this network, the human being has an important position, because we are the living beings in which the network reaches self-knowledge: nature knows about itself by our knowing about it. The human being is the only living being known to us so far in which nature or the universe recognizes itself.

Another obvious way to question human hubris, which is far from being overcome by evolutionary theory alone, is to put oneself in the position of living beings that have different or more finely adapted sensory organs. From the point of view of a dolphin, a bat, or even a rat, humans are pretty pathetic creatures – they can't fly or swim properly, they don't communicate by sonar, they have no stable community and they have a poor sense of smell. Who knows if our pets, especially dogs and cats, don't even take pity on us and bring us sticks or a mouse now and then, which they have hunted at night on their forays and adventures. Other creatures locate prey or members of their species via levels of reality that have become accessible to us at most through the technological extension of our senses. If it ever occurred to them to compare themselves with us, they could become victims of the same self-delusion and hubris from which humans suffer. Yes, they could even harbor **speciesist prejudices**, believing that their species is far superior to the very limited humans because of their peculiarities (e.g. a particularly good sense of sight in birds or the sense of sonar in bats), even more so because these humans, while simultaneously proud of being particularly intelligent, are so stupid as to pollute and destroy their own environment.

It should be considered as the really radical insight into the evolutionarily reconstructible origin of the species that nature is not hierarchically organized: the life forms which temporally precede the existence of the human being do not stand in any sense in a line with the human being and culminate more or less contingently in us. In the colorful network of life we are just as much a marginal appearance as in the structure of the universe describable by astrophysics. Our position in the Milky Way is at some random point in any one of innumerable galaxies, particularly distinguished from our point of view only because we find ourselves here. Strictly speaking, this is also an exaggeration, because the terminology of center and periphery has no meaningful application in the universe or in evolution. We are, thus, neither at the center nor at the margins of the universe and evolution, as these notions do not apply to nature! Thus the assumption that we're only a peripheral animal on a marginal planet says more about us than about our factual position in the universe, as we will see in the second part of the book.

The idea of a family tree encompassing all species, starting from very simple living beings and branching out more or less distinctly, obscures the radical insight of the modern life sciences that the planet is teeming with life on all levels. From a biological point of view, the innumerable forms of life are not arranged according to a hierarchy; there is no purposeful development bottom-up or top-down. In addition, we have by no means discovered all species and are miles away from understanding exactly how all these life forms are interconnected in evolutionary terms. This is precisely an important insight of the modern life sciences, which know more and more in detail but have to realize that they cannot gain an overview of life as a whole.

Every single human being is, strictly speaking, a whole zoo of living beings symbiotically fused in our organism – they live together in such a way that they all benefit from this in their survival. Without this coexistence, without symbiosis, we would not exist. The microbiome of a human being, i.e. the viruses, bacteria and fungi that colonize us, comprises considerably more cells than a human being without a microbiome, who could not survive without this connection with other living beings.

We are thus part of nature in a far more radical sense than is evident to us on a daily basis. For in our everyday life as minded beings we strive to domesticate our own corporeality by constructing many cultural layers (from clothing to higher forms of art, from organized small groups such as families to states and societies), thanks to which we seem to keep nature (both in and outside us) at a distance and under control.[33]

Continuity, Discontinuity, or Somehow Both?

Although most people think of their position in nature in terms of evolutionary biology, many continue to believe that they simply aren't animals and that they actually don't belong in the animal kingdom at all – due to reason, intellect, mind, or some other special faculty that actually makes us humans. Above all, modern creationists (who believe that God created the universe along with all living creatures only a few thousand years ago), with their unjustified rejection of evolutionary theory as a whole, repudiate the indeed unacceptably simplified notion that humans are descended from apes or from an ancestor that we share with other primates. But independently of the dispute between those who want to recognize a trace of divine intention in the blueprints of life and those who believe that all forms of life are only random mutations that can ultimately be exhaustively explained by means of evolutionary mechanisms, it is still not settled how exactly we fit into nature. One really does not have to be a creationist to realize that our mental faculties cannot be meaningfully explained in the language of evolutionary biology or evolutionary psychology.[34] In order to realize that evolutionary theory does not provide omniscience about the human being, it is sufficient to recognize that the humanities and social sciences as well as our human self-knowledge provide us with just as much information about the human being as modern life-scientific explanations.

In reference to the relationship between human beings and animals, one can to this day distinguish two rough traditions of thought: the continuity thesis and the discontinuity thesis.

---

*The **continuity thesis** states that humans are animals in the same sense as all other animals and that every characteristic that supposedly distinguishes us from other animals (morality, language, culture, self-consciousness, upright posture, or whatever else has been mentioned) can be observed in a different form elsewhere in the animal kingdom. If they differ at all, humans differ from other animal species by* gradation *but not in principle. The **discontinuity thesis**, on the other hand, claims that humans differ from other animals in principle by a number of characteristics (by language, reason, higher morals, science, technology, or whatever).*

The discontinuity thesis admittedly places the human being in the animal kingdom as well but leaves room for our own kingdom, which transcends animality, on the basis of specific, mostly intellectually determined abilities. This then raises the tricky question of how our higher intellectual faculties are connected with our animality. This question is not easy to answer for discontinuity theorists, but nor is it easy to answer for continuity theorists, as we will see.

In this context, a widespread diagnosis of continuity theorists has it that the defining idea of discontinuity theorists – that the human being is something special, a living being radically different from animals – is one major reason for our devastating tendency to exterminate other life forms and destroy 'the environment.' From this perspective, religious people who do not want to come to terms with their animality are often reproached that monotheism, with its conception of the human being as the crown of creation, has caused terrible damage to the other animals.[35] But this accusation is rather absurd if one considers that, in its long phases of religiously based societies, humanity has dealt with the environment in a considerably more sustainable and even respectful way than has been the case since industrialization. And when industrialization began about 250 years ago, the religious explanations of the world simultaneously lost significance and were replaced by a technocratic worldview – a process which has entangled us in a devastating climate crisis at the world-historically rapid pace of now a little more than two hundred years. In its coupling with an economic system based on the illusion of limitless growth, this modern worldview has caused more damage to the environment than all religions combined (which is not to deny that all religions – including the seemingly peace-loving Buddhism and Hinduism – have been producing violence and delusion). Unleashing the powers of industrial capitalism in the context of more or less secular societies, not the religiously motivated illusion that we are in no sense part of the animal kingdom, is the driving force of self-extinction.

In any case, industrialization was driven not by the fact that monotheists consider themselves to be the crown of creation but, rather, by the fact that scientific-technological progress was decoupled from the question of "the human place in the cosmos," as the philosopher Max Scheler put it.[36] Our interest in ourselves as humans was replaced by a machine age which promised to deliver economic and social flourishing regardless of how we see ourselves and our place in reality. The monotheistic representations of the relationship between the human being, the animal and God are responsible neither for industrialization nor for the fact that in modernity we have reached a crisis of sustainability on all levels of human coexistence. If at all, such a criticism applies to the worldviews associated with industrial capitalism, namely **naturalism** and **scientism**, which are based on the continuity theory, and according to which the world or reality as a whole can be completely explained and controlled by understanding the human being as a natural phenomenon, an element of a gigantic material-energetic system in which subsystems of different complexity (from elementary particles to solids to galaxies and the universe as a whole) interact. Naturalism and scientism believe that we can (in principle, not in reality) fully explain, predict and control the unfolding of processes on our planet by way of science and technology. If we look at the historical realities that are associated with this worldview, it turns out to be more dangerous than religious delusion – though we should, of course, not ignore the fact that modern science and technology are actually combined with religious fanaticism in those societies where religious worldviews dominate the fundamental structure of society. If one combines the naturalistic worldview with the management perspective of behavioral science, according to which societies could and should be controlled on a purely economic level in order to generate wealth that can be measured economically, you get the formula for modern self-destruction.

The modern naturalistic worldview communicates its vision in terms of what the German philosopher Hans Blumenberg has called "absolute metaphors." An absolute metaphor is designed to give us an overview of how a worldview conceives of humans in their relation to nature, the universe, animals and society. Paradoxically, the naturalistic worldview is in dire need of metaphors because it cannot actually deliver on its promise to explain everything in scientific terms. The most prominent metaphor is still the idea of the world (and everything in it, including ourselves) as a gigantic machine – or, more recently, as a computer.[37] This metaphor is not only a one-sided distortion of reality. It also contributes to the emptying of meaning from our concepts of human, nature, life, animal and environment. Industrialization, as a side effect, generated an experience of reality as devoid of meaning. Without such a conception it would be much harder to subject our labor forces and economic activities to the idea of extraction of resources at the service of economic surplus value production. In short, the naturalistic worldview and its nihilistic conception of reality as in itself a meaningless object of scientific investigation and technological control is part of the explanation why the industrialized nations are obviously not yet prepared fundamentally to rethink their way of life and to recouple scientific and technological progress with human and thus also moral progress.

Against this trend, some economists, such as Dennis Snower and Katharina Lima de Miranda, have developed new indicators for recoupling wealth and social well-being against which the social and ecological consequences of economic activity can be measured (in addition to gross domestic product, now seen as problematic by many).[38] And the aforementioned Tyson Yunkaporta – to mention just a few developments – is working in Australia with schoolchildren and students on the question of how indigenous knowledge could be used to correct the imbalances in our monetary system. If one sees progress exclusively in collecting individual scientific findings in order to make them technologically usable, one overlooks the ethical-philosophical dimension of human agency. To be sure, in order to know who we are, we need as much scientific perspective and factual knowledge as possible, but from such facts alone we can never infer who we want to be, let alone who we should be. Nature as an object of scientific modeling is not a source of values; it does not prescribe what we ought to do. While scientific research into ourselves as animals has been providing us with insights into our basic prosocial behavior (always also oriented towards the welfare of others), ethical insight can neither be justified nor derived from this factual knowledge. For systems of evil (such as the administration of a brutal dictatorship) also require prosocial, cooperative behavior. The factual structure of animal cooperation alone is not the ground of ethics.

In the following, I propose a counter-position to the view that the human being is an animal seamlessly joined with the rest of nature. For the idea that the human being is an animal in the same sense as all the other non-human creatures we consider as 'animals' today is one reason why we have adopted the modern attitude of dominance and control towards nature that has led us to the brink of self-extinction.

Shadowboxing

The concept of an animal and thus our understanding of our own animality (and thereby that of other living beings) on closer inspection turns out to rely on a human projection structure. We create the idea of a unified animal kingdom on the basis of a human, all-too-human representation of the animal, which can be seen through as a largely indirect expression of a disturbed relationship between ourselves and our own 'animality.'

*This then is the **projection thesis**: the human being makes an image of itself in which being an animal is the concept of a shortcoming which must be supplemented by a human particularity (the specific something). On the basis of such a concept, the human can only conceive of the other animals as humans first and foremost lacking the specific something. Animals thereby come into view as deficient beings. But precisely in this way one misses the particularity of the other* *living beings. While they have their own specific somethings, their animality is and remains a shortcoming they need to compensate by some feature or other. In this context, humans have created the idea of an all-encompassing environment, nature, which, in reality, is a large-scale projection of the human niche, our environment.*

Beneath the surface of this projection lies an awareness that we find in nature a gigantic range of life forms that are in fact deeply alien to us, which we bring closer and make more comprehensible by grouping them under the problematic concept of the animal as a deficient being (not quite or not yet human; less developed earlier stages of evolution; etc.). This then first raises the question how we can fit into this animal kingdom – a question that is misleading because this supposedly unified animal kingdom does not exist independently of our human projection. On this level our confrontation with the (other) animals and our shared environment, nature, is a kind of *shadowboxing*.

Of course, there are a large number of life forms. We are far from having discovered and classified all of them. We can only estimate how colorful life is on our planet; much is hidden from us because it is difficult for our measuring instruments to access it. The manifold forms of life cannot simply be divided into plants, animals and humans, because the non-human animals are as different from one another as we are from them. The concept 'animal' overlooks this insofar as it is so formulated that the animals become a kind of indistinguishable mass from which we humans differ by some specific characteristic or by a few such characteristics. It does not really help to add here that the animals also differ among themselves as long as we use the deficient concept of an animal to group the manifold life forms together.

It is typical in our approach to animals that we are always focused on the difference to humans or on the similarity to us: chimpanzees are close to us, bonobos even closer; we empathize with mammals – when we are not turning them into sausages – with amphibians less so, and even most animal rights activists kill insects without a guilty conscience. After a referendum on rights for non-human primates failed in February 2021, Swiss radio and television commented almost cynically on the negative outcome by saying that the primate initiative was "an interesting philosophical question – nothing more,"[39] which, in view of the massive suffering we humans inflict on other living beings, shows an ethically alarmingly low level of reflection. What we do to other living creatures (from factory farming to often brutal animal experiments or mass murders of living beings, e.g. minks in the recent pandemic in order to protect us from danger) is certainly not ethically justifiable. That we act in a radically evil way towards other living beings is an uncomfortable fact, which should not be glossed over by trivializing our cruel crimes as "an interesting philosophical question." The human being continues to behave as if we were the measure of all things, as the Greek thinkers Xenophanes and Protagoras realized thousands of years ago.

We do not, however, know what exactly animal ethics requires of us, though of course we know very well that mass factory farming in its present industrialized form, to choose the most obvious example, is morally reprehensible, and even radically evil.

Granted, no one defends an ethics of athlete's foot, which we fight without a guilty conscience with appropriate medication, because our animal ethics (as important as it is in view of today's ethically reprehensible factory farming) is limited to forms of life that sufficiently resemble our self-image as animals. Through the projection structure built into the concept of an animal, we make the animals, so to speak, images of our own animality, in which a theological heritage of monotheism certainly also resonates. After all, Noah does not take the bacteria, fungi and life forms capable of parthenogenesis, unisexual reproduction, with him on his ark … moreover, he does not and cannot take fish or other life forms which thrive in the water. Noah's ark is, therefore, a paradigm case of the projection thesis at work. This is not an objection to animal ethics but, rather, an indication that, in its present form, human, all-too-human projections are baked into it.

If we assume that there are different kinds of animals, this implies that there is a genus called "animal" such that the different animals then differ from each other through specific characteristics. But there is no criterion for when something is an animal, the characteristics of which are shared by all animals that exist, so that we could first recognize in all animals their generic, general animality and then specify it in detail. If zoology nowadays considers animals as multicellular life forms which are neither plants nor fungi (and are defined by their metabolism), unicellular organisms are excluded from the animal kingdom. Of course, it can be assumed that animals (which in zoology are usually called *Animalia* or *Metazoa*) are those multicellular organisms that eat other organisms, that can move, that develop embryonically in a certain way, and so on. Such determinations can be found in zoology handbooks, which deal with the question of how scientifically to distinguish "between NATURAL GROUPS or TAXA (taxa as products of nature and taxa in the sense of HUMAN CONSTRUCTS)." Thereby one learns that "the division into 'animals' including unicellular 'animals' and 'plants'" precisely does not reflect such a "systematic relationship," thus is a human construct. "On the other hand, it is assumed that all Eukarya are a monophyletic taxon, as are the multicellular animals (= Animalia or Metazoa)."[40] In the technical language of today's zoology, this means that all multicellular organisms go back to a primordial form from which they have evolved.

At the same time, we know that there are probably millions of as yet undiscovered life forms. Our taxonomy, i.e. our classification of life forms, is scientifically most secure where one can determine correlations by means of modern molecular methods (i.e. essentially genetics). But all this concerns first of all our classificatory interests and does not show that all living beings that exist in general are really objectively classified in such a way that they should be considered either as plants, as fungi, or as multicellular animals. The hypothesis that multicellular organisms are interrelated is currently considered validated, but it is of course a fallible scientific hypothesis.

Of course, the life sciences, including zoology, found out more about 'animals' than any other prior engagement with 'nature.' And most of those findings have only taken place since the second half of the last century (remember: Darwin could not have known the first thing about genes, as they had not been discovered back in his day). On the level of analysis of the life sciences one recognizes innumerable causal systems (blood circulation, digestion, etc.) interwoven in a complex organism such as a human body. And all the life forms with which we're familiar are in turn part of ecological systems consisting of organisms of different levels of complexity. The more the life sciences advance, the better we are able to explain how the many subsystems we find in organisms interact. Thanks to modern scientific advances, we know that there is a wide variety of life forms, from bacteria to elephants, from algae to bonobos, and so on.

Thanks to the microbiological findings of modern biology, we can now understand living organisms as complex systems that resist their dissolution into the environment, and so entropy, by means of certain biochemical and thus always also physically describable processes.[41] Therefore, it is natural to understand life and thus living beings via observable processes of self-preservation and self-reproduction (including growth by cell division, which is characteristic of complex organisms) and to define further criteria for life such as metabolism and information exchange with an environment.

And yet: as every textbook of modern biology concedes, we do not have a *definition* of "life" but only "practical LISTS OF PROPERTIES" of life, which include metabolism, low entropy, growth, reproduction and information processing. However, these are properties that "also occur in nonliving systems,"[42] such as combustion processes, crystals, computer viruses, and technical systems such as cell phone cameras.[43] The working conception of life that drives the life sciences is obviously sufficient for its discoveries, but it is far from telling us what life is or what animals are. Rather, hidden conceptions of life (and its meaning or meaninglessness) drive our research into the complex living systems to be found on our planet.

In defining what life is in the first place, we quickly encounter border areas and zones of not-knowing. For example, it is not easy to answer the question whether viruses are alive and how living systems evolved from non-living matter in the first place. During the years of the coronavirus pandemic I had the chance of talking to many life scientists about the ontology of viruses. They all explained that the exclusion of viruses from the realm of living matter is at best plausible by the arbitrary determination that viruses can only reproduce in a host. But this is no proof that they are not alive, only an insistence on a certain conception of life (having to do with autonomous reproduction). In any case, since viruses can be studied in evolutionary terms and thus by the life sciences, it seems obvious to include them in the realm of the living, as some renowned virologists and immunologists conceded in these discussions.

If it is questionable, as suggested above, whether the criteria used for living systems do not also apply to systems which at first sight do not seem to be living beings and which do not consist of organic matter at all, then why are those systems considered to be living, but not those that fulfill the same criteria (e.g. crystals or even stones) but which we do not want to classify as living? If one does not have a definition of "life" but only lists of criteria that are sufficient for lab work, this cannot be answered. This raises the suspicion that, despite all scientific progress in the life sciences, we currently do not really know what life is.

Consider simulations of the living such as the computer game *Game of Life*, developed by the mathematician John Horton Conway in the 1970s. Its rules are based on a theory of cellular automata that are subject to certain mathematical patterns, leading players to believe that living patterns are emerging before their eyes from simple shapes.[44] Therefore some scientists, such as the MIT physicist Max Tegmark, argue that life is not even essentially tied to organic matter but, rather, to formal patterns of self-replication (the repetition of similar patterns, e.g. through cell division), thus blurring the supposedly obvious boundary between inorganic natural phenomena such as meteors on the one hand and plants and animals on the other, at a high level of abstraction.[45] Tegmark thus raises the question whether systems of *artificial* life that can be programmed by us and implemented in non-organic hardware are alive, i.e. whether a certain form of (organic) matter is necessary for life at all.

Against all of this, I have argued elsewhere that life essentially had to develop over millions of years and that it needs a suitable material basis (including carbon and other molecules originating from imploded stars).[46] Regardless of who gets this right, these fundamental questions cannot be answered simply by referring to life-scientific findings and lists of characteristics of the living. The question of what life is cannot be answered exclusively by the natural and life sciences, an insight that is at the center of recent discussions in the philosophy of life.[47]

To sum up: the life sciences have long since ceased to operate with the idea that reality can be divided, so to speak, into non-living objects on the one hand and plants and animals on the other hand. In our everyday lives most of us hold on to this division from our human perspective and are surprised when we learn at some point that it is by no means so simple. Yet, despite their technical sophistication, and some useful and impressive research, the life sciences cannot really tell us what life is. And, to the extent to which they still sometimes draw on the deficient notion of an 'animal,' they need a philosophical update (just as much as some contemporary philosophy of life needs to delve deeper into the gray zones of the life sciences in order to overcome the idea that life is organized as a kind of evolutionary hierarchy with lower and higher stages).

What Does it Actually Mean to Understand Oneself as an Animal?

So far I have argued that it is a mistake to understand humans in contrast to animals in order then to pass judgment on how far the difference extends and in what exactly it consists. Rather, on the basis of the apparent obviousness that we are animals, we should first understand that we are therefore far from being *just* animals. Strictly speaking, no single form of life that we find is *only* an animal, but every form of life is different from every other. Of course, we can find patterns in life science, which describe, e.g., reproductive probabilities and routes without being able to derive a catalog of criteria from it which makes a human being a human being (for example) by the fact that he can only produce offspring with human beings (which, as we have already seen, is the Aristotelian idea of the specification of species). But humans, donkeys and dolphins are not animals by the fact that they reproduce through sexual reproduction, first of all, because not all forms of life develop through sexual reproduction (some living beings can reproduce by parthenogenesis, e.g. aphids); and, secondly, because not all individuals of a supposedly uniform species are able or willing to reproduce anyway. Just think of bees, where only a few individuals maintain the sexual reproduction of the collective. Abstract patterns of speciation that apply to some individuals of a given species or other fully cover neither the species (as they leave out some of the individuals) nor the concept of an animal.

If now the human being is an animal or, as I might say, the animal *par excellence*, our self-determination as an animal of a certain species directly or indirectly articulates the connection of our animality with our humanity, through which we obviously differ massively from other living beings. But thereby the difference between humans and other life forms is no longer reconstructed along a difference between being human and being animal. What we share with other living beings is being alive, not being an animal. What differentiates us from other living beings is not a specific something but an indefinitely long list of activities and biological features, which is why it is rather easy to identify a human when one meets one in contradistinction to other life forms. It is not difficult to distinguish a human from a bonobo or a bonobo from a bee. This is not to reduce living beings to their phenotypical appearance. I am just insisting on the fact that the realm of the living is sufficiently complex so as to generate forms that cannot be captured in terms of a more or less simple and clear-cut tree of life. Here, as always, reality is more complex than our concepts make it seem. Through a rhizomatic conception of life as exceeding our classificatory expectations, the other forms of life are freed from the yoke of being animals, in that we no longer understand them as *mere* (for instance, instinct-driven) animals as distinct from the specific human animal. Those life forms that we still address as animals are thus no longer seen as deficient human beings who have metabolism and instinct but lack intellect and language, say. So, the so-called non-human animals can be freed from their too close connection to our self-image as animals, and they can be recognized in their own right.

The question to which both the continuity and the discontinuity theses offer answers (Is the human being different from the animal by gradation or in principle?) therefore proves to be defective, as it presupposes that there exist such a thing as *the* animals to which humanity either belongs or does not belong.

The fact that zoology creates systematic classifications of metazoa (i.e. multicellular, complex living systems) through precise observation of natural phenomena, coupled with modern life science methods and knowledge, does not mean that there is an animal kingdom to which humans either clearly belong or clearly do not belong. Thanks to the findings of modern life science, zoological classification today no longer relies on the concept of animals (animalia) but, rather, can divide organic tissue forms into patterns at the elementary cellular-biological level. The classification of natural events into living and non-living no longer needs the concept of animals.

Thus, humans share many characteristics with other living beings on an organic level, but they also differ organically and by their manifold abilities from many other living beings, without our being able to derive a definition in the classical sense from this. Moreover, we are special in so many ways, i.e. different from other living beings, that it should actually be absurd to look for a single specific characteristic or a small set of characteristics that should distinguish humans from other living beings. One does not need extensive scientific studies to determine that on our planet only humans do mathematics as science, make movies, buy bathing suits, become committed to feminism, or stand up for animal protection. Just as cats and bats do things that would not be possible for us.

Just as it is in fact obvious that we are animals in some sense, so it is in fact obvious that we are distinguished by an "abyss," as the French philosopher Jacques Derrida has rightly said, from the other forms of life that we sometimes like to call and consider as "the animals" as distinct from humans.[48] In sum, neither the continuity thesis (which works with an unclear conception of animality) nor the discontinuity thesis (which suffers from the same defect) captures the essence of being human.

If the concept 'animal' arises from the fact that the human being regards itself as an animal and then projects this being-animal onto other living beings, as I have argued so far, one can go so far as to claim that the human being is the paradigmatic animal – that is, the model for how we imagine the other animals.

---

*Let us call this the **first fundamental anthropological principle**: our concept 'animal' is always designed in view of humanity, from whom we take our starting point in our investigation of the animal kingdom. What appears to us as animal, we consciously or unconsciously take from our own nature and then project this more or less uncritically onto the other forms of life. In this way we erect a distorting mirror in which we no longer recognize ourselves and the other living beings correctly. In place of the question of how humanity relates to the animal, we should therefore ask how humans determine themselves as animal.*

The continuity and the discontinuity theses are also both partly true, even if they are based on an ultimately mistaken question: insofar as we are alive, there is indeed a continuity with other life forms. This continuity can be investigated better than ever before with the appropriate methods of the modern life sciences. We now know more about humans as life forms than all of humanity did before the twentieth century. That is impressive. But exactly this impressive scientific achievement is direct evidence for the discontinuity thesis, because obviously only humans are interested in and capable of developing scientific theories about humans as animals as well as about the building blocks of life, which we share with all forms of life known to us.

The human being is accordingly the only known living being which is clearly interested in comparing itself with other animals in order simultaneously radically to differentiate itself from them. The human being is therefore the animal that, starting from its self-observation, develops a concept of animal and then imposes it on the other forms of life. In this way they appear to us as defective, as deficient beings, because they are like human beings (i.e. animals) who lack something, namely that by which we define our special position.[49]

---

*This leads to the **second fundamental anthropological principle**, which is that the human being is the animal that does not want to be one.[50] By establishing our own 'animality,' we are at the same time anxious to strip it off and to identify ourselves with that part of ourselves by which we differ from the other animals. We want to be special.*

In this context, it turns out that, from the human point of view, the other animals traditionally merge into a kind of biomass or a large undifferentiated clump of animals. In the end it is nonsense to distinguish the forms of the living into human beings on the one hand and animals on the other hand. Mind you, it is just as nonsensical to classify humanity into a so-called animal kingdom, as it were. There is a variety of cellular patterns and elementary forms of life with which our planet is teeming, and which are not yet known to anyone in their entirety. But, to repeat this crucial point, there is no animal kingdom which consists of tree-like structured lineages – an ultimately ancient way of thinking which we have not completely abandoned to this day.

The reason why we should distinguish humans from non-human animals should actually go hand in hand with the realization that the many organisms and life forms that we find apart from humans – from bacteria to mayflies to lapdogs – must not be reduced to being mere animals, i.e. instinct-driven products of evolution, any more than humans should. It follows that, strictly speaking, no animal is *just* an animal, because the very idea of being an animal is a false abstraction and projection of our own split self-relationship.

Insofar as the human being is the paradigmatic animal (that is, the living being from which we derive the concept of animal), we can study in ourselves to what imbalances, forms of decay and pathologies we give rise by conceiving ourselves according to the pattern of an undifferentiated concept of animal and thus projecting an ultimately disturbed self-image onto the other animals, which are in truth deeply alien to us.

---

*Somewhat paradoxically formulated, one can thus venture a **third fundamental anthropological principle**: the human being is the only animal known to us. Our concept of an animal is a self-conception of a part of ourselves, the part that is left over after we have identified ourselves with our most cherished special something (language, culture, morality, history, technology, or what have you).*

Instead of the idea that our humanity is to be located in the animal kingdom, I would like to argue the opposite view, namely that the animal kingdom is to be discovered in us, in our biological structure, in our thoughts and self-images. The aim of the following anthropological self-exploration is in the first place to recognize the strangeness of the other 'animals' in order then to realize in the following two parts of the book that we ourselves are just as strange because of our paradox-ridden animality.

Nature is not only 'out there,' 'in the wilderness'; it is not a 'foreign body' which has to be explained and controlled by science and technology. Rather, nature in and outside of us is essentially something that we can only ever take note of inadequately and incompletely. The phenomena of our animality, to which we count our desires, our feelings, our vulnerability and our mortality, reach into the unknown essence of the human being. This should be the source of an appropriate humility in the face of the almost infinite superiority of a profoundly alien nature over our quite impressive ability partially to understand and control this alien nature in and outside of us. But this understanding and this control make us neither masters nor mistresses of creation but, rather, participants in a complex event that ultimately always exceeds our power of comprehension.

Why We are Not Amphibians

The complexity of our situation does not go away when we turn to the other side of the coin of our self-image as rational animals, i.e. to reason. As Mephistopheles states in Goethe's *Faust* in his dialogue with God in the "Prologue in Heaven":

> Earth's little gods still do not change a bit,
> are just as odd as on their primal day.
> Their lives would be a little easier
> if You'd let them glimpse the light of heaven –
> they call it Reason and employ it only
> to be more bestial than any beast.[51]

Mephistopheles formulates a deep insight here, even if he does not mean it quite so seriously. This insight has it that the idea that in addition to our animality we also have a specific difference with respect to the animal, i.e. reason, by no means automatically elevates us above nature. Thus Mephistopheles (and on this point Goethe must have agreed with his fictional character) rejects the amphibian theory, as I call it.

*The **amphibian theory** (a variant of the discontinuity thesis) says that humans lead two forms of life (from Greek amphi-bios = double-lived). On one side we are animals like all other animals; on the other side we live the higher life of reason.*

It fits with this theory, which goes back to the ancient Greeks, that Aristotle could rely at the time on the fact that there are at least two word groups for life in ancient Greek: *bios* and *zoê*. The philological and philosophical aspects of this distinction are intricate.[52] Nevertheless, it can be generally stated that even the supreme God, whom Aristotle envisioned as the unmoved mover of the cosmos to which everything is oriented, is a living being (*zôon*), or an animal for that matter. This does not mean that God should be straightforwardly conceived as an animal; rather, it should mark out one side of our life, namely that which actually or allegedly brings us into contact with the divine. Mephistopheles calls this – with intentional ambiguity – the "light of heaven." For Mephistopheles, the human being is ultimately only an animal, which is why he takes Doctor Faust to task, to prove against God that even the modern pure spirit of inquiry is driven by worldly desire.

Admittedly Mephistopheles loses his bet at the end of *Faust: Part II*, in which it becomes clear that the heavenly and the human-sexual are interwoven. Nevertheless, Mephistopheles is right on one decisive point: the human use of reason does not overwrite our animality but, rather, transforms it.

Philosophical anthropology is concerned with the question of who or what the human being is. Every concrete answer to this main question of anthropology provides a conception of humanity. In general, a conception of the human being is a self-determination of the human being, i.e. a concrete conception of that wherein being human actually consists.

The amphibian theory is one influential conception of being human, one out of many pictures of the human being. This picture of the human being can be found in many traditions. But it is expressed particularly clearly in Plato and Aristotle, who distinguish three kinds of aliveness or being ensouled, each with a slightly different accent.

In Plato, the faculty of desire (*to epithymêtikon*) stands in the lowest place. This desire expresses itself in our life as elementary striving for pleasure and avoidance of displeasure. Paradigmatically it is bound to food intake and sexual reproduction. Over the millennia, from this faculty developed both the theory of the unconscious and, in its present form, the study of elementary animal impulses that we share with other animals. The middle rank is occupied by *thymoeides*, by which Plato refers to the impulses of excitement and indignation. The *thymoeides* is that dimension of our life, thinking and feeling that today is paradigmatically claimed by the social networks and online media. The middle level of the human animal is that which is exposed to manipulation, ideology and propaganda by being carried away from one opinion to the next without hesitation.

The highest point of human life transcends the animalistic processes of desire and impulsive opinion. Plato speaks of the *logistikon* – that is, the rational part of ourselves. In the wake of Plato, antiquity was occupied for centuries with the question whether the rational part of a human being is immortal or not – a topic that should not be put aside prematurely. For from Plato to the modern enlightenment of Moses Mendelssohn and beyond, serious philosophical reflections were undertaken on the immortality of the soul.[53] But that is another matter and will occupy us further elsewhere (see pp. 119ff.).

It is important that Plato, as well as his student Aristotle, and subsequently several thousand years of European and non-European intellectual history have supported an amphibian theory.[54] According to this theory, the human being is on the one hand an animal like all others, but on the other hand a special animal. The characteristic of the human being – our specific, anthropological difference – in part elevates us above the animal kingdom and connects us with the dimension of the divine. Thus, the human being lives at least two lives, the life of an animal and the life of a godlike being, whose immortality can then be discussed in further detail. Yet, this leads immediately to a worry which became a prominent objection in contemporary philosophy of the animal.

The **objection of two lives** states that a human being cannot lead two lives at once, the life of a specimen of the animal species Homo sapiens *and the life of a thinking person who is never entirely identical with his body.*[55]

Notice that the amphibian theory does not rely at all on monotheistic ideas of the soul (prominently developed in Christianity and Islam). And one finds it by no means only in European intellectual history. Wherever the belief appears that the human being can be reborn or ascend into other spheres of existence, as in ancient Egypt, in Hinduism or in Buddhism, one encounters elements of amphibian thinking. Like many ideas nowadays often associated with European thought, the amphibian theory predates anything recognizable as Europe.

The Animal Word – Why the Zoo Does Not Exist

In our century, the amphibian theory has been put in question by some schools of thought in animal philosophy and anthropology. In this connection the animal philosophy of the French philosopher Jacques Derrida is of utmost significance, a philosophy which in many respects ties in with the extensive works of the French philosopher Élisabeth de Fontenay.[56]

In his late work *The Animal That Therefore I Am* (2010), Derrida developed the idea that the concept of the animal radically distorts the reality of the manifold forms of life that are called "the animals." For everything one puts together under the term of "animal" leads inevitably to a confusion, which he calls "chimera." Here, a chimera is the name for an illusion, which is an allusion to Greek mythology, in which the Chímaira is a fire-breathing mixed creature mentioned in the archaic epics of Hesiod and Homer. The Chímaira is a fusion of lion, goat and dragon and thus one of the most famous mixed creatures in history.

Derrida now elaborates that the concept of animal actually refers to such a hybrid being, to a colorful animal kingdom, which in reality consists of so many forms of life that it ultimately eludes our efforts to classify it.

> *Ecce animot.* Neither a species nor a gender nor an individual, it is an irreducible living multiplicity of mortals, and rather than a double clone or a portmanteau word, a sort of monstrous hybrid, a chimera waiting to be put to death by its Bellerophon.[57]

Derrida recognized that the concept of the animal is an expression of an "animality of certain animals" [*animalité de certains animaux*],[58] ultimately a projection of our own animality onto the animal kingdom, which in reality is and remains deeply alien to us. But Derrida is not suggesting that since Darwin we have learned to place ourselves in the animal kingdom. On the contrary, he is concerned with emphasizing the radical heterogeneity of animals on the basis of an "attention to difference, to differences, to heterogeneities and abyssal ruptures as against the homogeneous and the continuous":[59] "I have thus never believed in some homogeneous continuity between what calls *itself* man and what *he* calls the animal."[60]

---

*Regardless of our classification of the animal kingdom, which until the twentieth century was predominantly made on the basis of phenotypes because genetics was not yet in possession of the technological possibilities available today, there is – according to **Derrida's animal-philosophical insight** – nothing in reality that corresponds to our confused concept of animal.*

Derrida articulated this idea in a series of puns. The most important one in our context is *l'animot*. When pronounced, this word sounds like the French plural of *animal* = animal, i.e. *animaux*. The word consists of *ani* and *mot*, the French word for "word." With this, Derrida refers to his philosophical starting position of **the linguistic turn**, according to which the method of philosophy consists in analyzing language and thus the signs by means of which we make reality intelligible to ourselves in the first place. And it is within this framework that Derrida comes to the realization that it is no coincidence that it is through its self-designation as a rational and linguistically gifted living being that humanity distinguishes itself from other life forms, which it melts down into a chimera or a bioclump, culminating in the questionable concept of "biomass."

Let's call **zoo theory** the idea that (as in a zoo) we put the life forms we encounter into enclosures according to our convenience and categories, which include the zoological classifications that turn the biomass back into a variety of different 'animals.' The zoo does not exist independently of our setting it up – and there is a separate historical field of research that deals with the question of how zoos were actually introduced in the wake of the modern scientific revolution and the Enlightenment in order to provide a visible sign of successful domination of nature within human cities.

However, there's one thing humanity hasn't reckoned with. Animals cannot be contained, because the diverse life forms we know today are by no means found only "out there," in the "wilderness," but also inhabit our bodies and are always to be found around as well as inside us. Humans do not live in a secured area, where viruses, fungi and bacteria invade from time to time. Rather, there are microorganisms and small living beings everywhere, without which we could not survive. Humans exist in many symbioses; our life requires that we are intertwined with other living beings and, at the same time, that we kill them with almost every breath.

Animalism, The Prestige *and* The Anomaly

For a number of years now, the expression "Animalism" has been used in philosophy as the counter-thesis to amphibian theory.[61] If we follow the definition of the term by Stephan Blatti and Paul F. Snowdon, who published a seminal anthology on animalism in 2016, we can first state:

Animalism *is "the claim that we, each of us, are identical to, are one and the same things as, an animal of a certain kind. That kind is what is called Homo sapiens. Putting it less technically, each of us is a human animal. According to this proposal we can say that at various places there is both an animal and one of us, and those things are in fact the same thing."*[62]

In contrast to the amphibian theory, animalism claims that we are animals through and through, that there is a specific difference between human and non-human animals, but not a difference between our animal side and our non-animal side. According to this view, the human being is completely and utterly an animal.

To some readers, animalism will seem to be such a matter of course that they can hardly imagine how one could seriously hold any other position. For this, one might want to rely on the simplest findings of modern biology since Darwin: insofar as the human being is part of evolution, it is of course an animal – what else would it be? But it is by no means that simple, as the critics of animalism show. The American philosopher Eric Olson (who himself defends a variety of animalism) sums it up when he writes:

> What sort of things might we be? Let us begin our study of answers to this question with the view that we are animals: biological organisms, members of the primate species Homo sapiens. This has a certain immediate attraction. We seem to be animals. When you eat or sleep or talk, a human animal eats, sleeps, or talks. When you look in a mirror, an animal looks back at you. Most ordinary people suppose that we are animals. … But things are not so simple. … the appearance that we are animals may owe merely to our relating in some intimate way to animals – to our having animal bodies, if you like – rather than to our actually being animals. The weight of authority is overwhelmingly opposed to our being animals. Almost every major figure in the history of Western philosophy denied it, from Plato and Augustine to Descartes, Leibniz, Locke, Berkeley, Hume, and Kant. (Aristotle and his followers are an important exception.) The view is no more popular in non-Western philosophy, and most of those writing about personal identity today either deny outright that we are animals or say things that are incompatible with it.[63]

Let us look at a prominent contemporary example of a critique of animalism, the eminent British philosopher Derek Parfit, who passed away in 2017, and who contributed a seminal essay to the aforementioned volume on animalism. Parfit became known beyond specialist circles for his thought experiments on personal identity, which can be illustrated by films such as *Star Trek* or the magnificent film *The Prestige* (2006) by Christopher Nolan.[64] The starting point of Parfit's critique of animalism is the question whether an exact, in every respect qualitatively indistinguishable copy of my organism, which arises at another point in the universe, is identical with me. This is precisely the case that occurs in *The Prestige*. In Nolan's film, a magician named Angier, with the help of an apparatus made by Nikola Tesla (played by David Bowie), develops a magic trick whereby a person who climbs into a tank of water drowns there in agony while an exact clone is created, which, after the death of the original, appears as the latter and reaps the glory. The clone is indistinguishable from the original, even from the inside view. It does not feel itself whether it is the clone or the original, because it is created with all the inner mental states of the original at the very moment the latter dies. Parfit plays this magic trick in his works intellectually with a so-called teletransporter, by means of which one can arise anew at any place of the universe – at the price that the original, who climbs into the transporter, is annihilated.

Hervé Le Tellier employs the same idea in his novel *The Anomaly*. In the book, the same Air France plane lands a second time in New York three months apart with the same passengers, so that the fellow passengers are suddenly doubled and meet each other in the course of the novel. The novel describes some of the episodes of the meeting of these paradoxical twins without offering a resolution of the question how the doubling came about in the first place. Regrettably, however, the novel comes down to the so-called simulation hypothesis, which goes back to the philosopher Nick Bostrom and which I have presented (and rejected) elsewhere.[65] According to this hypothesis, it is more likely that we are figures in a computer simulation that has produced a far more intelligent civilization of the future than that we are the flesh-and-blood beings that many believe themselves to be.

To be sure, this hypothesis raises the problem of our true identity: are we perhaps only sims in a simulation, are we animals, or are we identical with our streams of consciousness, so that it possibly plays a subordinate role whether we are carbon-based or stored on silicon chips?

*These simple thought experiments, cinematically and literarily appealing, sharply illustrate the **problem of temporal personal identity**. This consists in the fact that one cannot say without further ado whether and in which respect I am the same as I was twenty-five years ago. If all cells of my body were replaced by other cells, organized in the same way as mine (which actually happens for many body cells and whole organs such as the skin several times in life), would the resulting animal be identical with me or not?*

Animalism affirms this and argues that the continuity of my person, my self, over different stages of life is *biological*, which is why one speaks here of **biological continuity**. This differs from another approach, defended with particular prominence by Parfit, which sees personal identity over time in terms of **psychological continuity**: according to this view, I am the same as I was twenty-five years ago because there is a mental connection between my self today and my self back then. Relevant to this view is the work of the British philosopher John Locke, who understands a person to be "a thinking intelligent being, that has reason and reflection, and can consider itself as itself, the same thinking thing, in different times and places."[66] Even if one doesn't believe in either biological or psychological continuity, one cannot easily discover a substantial error in these assumptions. The reason why one can believe oneself to be something which is not identical with an animal is because we can imagine that the animalistic, outwardly visible shell of a human being lives on, while we are firmly convinced that the person has already died. Parfit recalls the hotly debated case of an accident victim, Nancy Cruzan, whose brain stem kept her body alive for many years while it was determined that her cerebrum had died and she therefore (presumably) no longer continued to exist as a person. In any event, on this basis, the US Supreme Court in 1990 allowed her parents to become entitled to discontinue artificial feeding of their daughter's mortal shell. Parfit comments on this ethically difficult case as follows:

> When Cruzan's cerebrum died, her parents came to believe Cruzan the person departed from her body, though the human animal continued to exist with its heart beating and its lungs breathing until, after the feeding tube was removed, the heart stopped and the animal was at peace.[67]

This consideration is an expression of a rather crude dualistic conception of an animal. If one imagines a Dalmatian to be brain-dead, but artificially breathing, it would still be a Dalmatian according to this logic, and not only its shell. However, many animals have consciousness, so that Parfit implicitly distinguishes here merely between living beings with consciousness and without consciousness, which, however, throws no light on the question to what extent humans are animals.

Another thought experiment that plays a decisive role in the discussion about the scope of animalism revolves around the question whether one would still be oneself if one's cerebrum were transplanted into another body (one can have justified doubts whether this would be possible at all; in any case, it is not clinically feasible so far). Here

Parfit decides that one would still be oneself if one woke up one day in the body of another person, from which he concludes that one is in no way identical with a whole animal but at most with a part of an animal (the cerebrum). But even against this, in his own view, it cannot be completely dismissed that one would still be oneself even if one spontaneously appeared in another place as an exact physical and psychological copy of oneself. But here too Parfit's consideration does not lead to a self-knowledge of the human form of life, because one could ask the same question as well on the basis of brain transplant and the complete copy of a Dalmatian.

Parfit distinguishes three conflicting strands of theory that should nevertheless be taken seriously: *immaterialism*, *Lockeanism* and *animalism*:

---

*Our personal identity across different time intervals (and social roles) accordingly consists in being an immaterial thinking substance (**immaterialism**), in believing ourselves to be the same person who recognizes themselves as themselves (**Lockeanism**), or that we are the same organism that for a time resists entropy, i.e. the dissolving tendency of structures in the physical universe (**animalism**).*

Besides the philosophical disputes separating these positions in the narrow sense, something else is of importance for our context. Both animalists and their opponents rely on a conception of animality that they don't question. While the animalists are of the opinion that modern biology has taught us that we can join the animal kingdom more or less unproblematically, the anti-animalists hold against this view that we are not merely animals of a certain kind, and thus objects of scientific research, but also persons with mental capacities. Since Locke, **person** is understood as a legal concept: a person is a self that has certain rights and duties that derive from the fact that we are endowed with reason.[68] A person is defined by social and moral relations and does not merge into the material-energetic processes of self-preservation by which an animal maintains itself as an organism against environmental pressure. Admittedly, not only humans but also other living beings have rights, which is an achievement of the constitutional state, thanks to which we can legally stop at least some unnecessary and morally reprehensible cruelties towards other living beings. But it is difficult to derive animal duties from these so-called animal rights and to ask dolphins or gorillas to improve gender equality in their societies.

The Human Animal as Machine?

Animalists and anti-animalists share the assumption that by an "animal" we should understand something that is the object of the life sciences. But the life sciences, on their part, are – for good methodological reasons – committed to mechanism, i.e. to a certain form of thinking. And this form of thinking reaches its limits when we try to identify ourselves as human beings with a complex cellular automaton. **Mechanism** as a form of thinking consists in understanding a system of elements (such as cells) arranged and interrelated in a certain way from the simplest possible interaction between its parts. Mechanism breaks down a complex living system, such as a whole human animal, into many subsystems, which in turn can be studied down to their smallest details, thus indirectly understanding and influencing the functioning of the system as a whole.

We owe modern human medicine to the mechanistic form of thinking and were recently able to develop a whole range of vaccines against SARS-CoV-2 with impressive speed. If you go to the doctor, it is obviously better that he or she has usually enjoyed a largely mechanistic training, first breaking the patient down into subsystems, for which there are then individual specialists who are hopefully up to date with the latest research. In this respect, there is nothing wrong with mechanism as a method. But this does not mean that mechanism represents a sensible or even correct view of the human being – only that mechanistic thinking produces important scientific methods and models.

Mechanism is opposed by **holism**, which rightly insists that a living being is ultimately an organism only as a whole. A heart beats as part of systems that work together. A heart beats as part of systems that, taken together, all lead to my being able to live a (hopefully long and healthy) life as a human being. The holistic approach understands the subsystems of the organism from the point of view of their functional position in a whole, which in explanatory logic precedes the parts. Especially in the life sciences, holistic and systemic thinking has recently played a serious role, as the renowned Oxford physiologist Denis Noble shows in his highly recommended books *The Music of Life* and *Dance to the Tune of Life*.[69]

The fact is that today we only understand some parts of the total human biological system, and this for a number of reasons. As a whole, the human being is simply too complex to be fully explained mechanistically. In this context, **complexity** means at least that we can only explain and understand a given whole by decomposing it into subsystems, whereby this decomposition in turn dissolves or changes essential properties of this whole system. Complex systems cannot be fully understood by decomposing them into parts that are better understood, because this decomposition changes essential properties of the system that we want to understand. Complex systems are not predictable; they never behave in a way such that we could totally control them.[70]

Here's a concrete example from the everyday reality of the coronavirus pandemic, which has drastically demonstrated the phenomenon of complexity: by contributing to the restriction of contact through the empirically sensible lockdown of subsystems of society (e.g. school closures), new contacts are created elsewhere. This is because people withdraw into private households in small groups, since complete isolation is not possible and would not anyway be desirable, especially for children. In the gray zones of the lockdowns, the virus then continues to spread in a way that is no longer politically controllable, so that the pandemic shifts to contact spaces in a weakened form. The complex system of the "pandemic" cannot therefore be simply broken down into the smallest possible units – individual infections or even individual virus particles – without changing it in an uncontrollable way. This is because a pandemic also involves a social reaction.

Paradoxically, measures to combat the pandemic generate new probabilities of infection, which can be contained by further restrictions on contact until a limit is reached – a limit which, at least in liberal democracies (fortunately!), cannot be moved any further without violating elementary human rights. The interventions in basic rights that were made to cope with the pandemic were already drastic. But even these interventions were not able to end the pandemic quickly, and they moreover produced incalculable collateral and consequential damage elsewhere in the maximally complex system of "society."

An overemphasis on mechanism in modern thought (especially since René Descartes) has infamously led to the view that animals are machines, systems that ultimately consist of many parts whose behavior is identical to the overall system. Descartes explicitly sees animals as collections of organs. Humans, according to Descartes, act "from knowledge," animals "from the dispositions of their organs."[71] The essential difference between us and the animals, according to him, is that while we have reason and language as a "universal instrument,"

> … which can serve for all contingencies, these organs have need of some special adaptation for every particular action. From this it follows that it is morally impossible that there should be sufficient diversity in any machine to allow it to act in all the events of life in the same way as our reason causes us to act.[72]

Descartes reduces animals to a collection of organs, an organism, which he understands as a machine, whose parts produce certain effects. That is why in this context one also speaks of **reductionism**, which basically means the reduction of a whole to its parts. For example, someone who explains the behavior of a human being completely by neuroscientific research into his neuron patterns thinks just as reductionistically as someone who attributes the identity of a person over different time intervals to the biological continuity of an organized cluster of cells, a human animal.

In the wake of Descartes, Julien Offray de La Mettrie – who was persecuted for his theses and found shelter in Potsdam with the philosopher king Frederick the Great – took a crowbar approach to humanity in the eighteenth century. In his pamphlet *Man a Machine* (*L'Homme-machine*), he argued unequivocally that all animals, thus the human being as well, are soulless machines.[73] His starting thesis is that the entire universe is only one gigantic machine which can be completely explained by the interaction of its parts, so that the human being as an animal is also completely absorbed in this structure. For La Mettrie, everything that exists is machine-like, explainable from the interaction of individual parts. Thus the animals, and so human beings, can be no exception.

La Mettrie is thus the key witness for the massive projection of an error about the human being onto reality as a whole. His error consists in mistaking his own animality for a gear wheel of small particles, which looks in the mirror like La Mettrie. He then transfers this distorted image of himself to the other animals, whereupon he performs a *salto mortale* in the last act and considers reality as a whole to be a soulless machine.

One of the most serious errors of mechanism and thus also of normal reductionism consists in a confusion of two kinds of totality – *sums* on the one hand and *wholes* on the other – which was recognized particularly clearly by the unjustly forgotten biologist and philosopher of life Hans Driesch, who was an important representative of holism.[74] A mere sum forms a totality, but not a whole. It is the case, as Driesch remarks, that a

> certain *unity* ... will certainly always exist with *natural objects*, which are sum-like entireties, insofar as the relation of mutual attraction, perhaps practically not provable, exists among their parts. There is therefore a "unit of action" there.[75]

What Driesch formulates here somewhat tortuously is the thought that processes and objects in the scientifically observable universe always interact with partly distant processes, because they belong to the same physical fields. My desk and my coffee cup interact with each other (the cup is just in front of me on the desk). Nevertheless, the two form a sum only because their unit of action (on my desk) can be dissolved without the objects ceasing to exist. It is different with a living being. If I cut a frog into two halves (please, don't do it!), it ceases to be a frog. A frog is therefore not a summary heap of cells but an organism.

In the summative view of nature everything is connected with something else to the extent to which there is some causal contact between parts. In this way, one finds in nature, so to speak, the most adventurous structural connections. From this perspective, even the tip of my nose forms a totality with the Eiffel Tower, because both are causally interconnected (I could touch the Eiffel Tower with the tip of my nose, for example). Also two persons form such a causal sum when they shake hands or kiss each other. But in this case the two persons do not merge completely.

We need another (not merely summative) additive concept of a totality, which Driesch describes as a *whole*. Wholes are "totalities, which impose another concept on the observation than that of the mere sum. For the time being, I name at random: table, frog, musical score, volitional content, meaning structure."[76] As a compelling example of the difference between sum and wholeness, Driesch cites the difference between the mere word sum "Because although dog with that without" and the sentence "Greece gave birth to many great thinkers." The first word sequence has no sense, even if it has an order given by the sum of its parts. The second word sequence, on the other hand, has a meaning and thus shows a unity which welds the parts of the sentence to a whole which is obviously not identical with a mere sum of its parts. This can also be described by the difference between two people who kiss, and thus merge into a unit of meaning for the duration of the kiss, and two people who happen to sit next to each other on a bus and do not form a unit of meaning.

Driesch now argues that organisms must be understood essentially from a "holistic causality,"[77] a view expressed in his much maligned (but hardly so much as understood) doctrine of vitalism – a tradition of thought that is being revisited in contemporary environmental philosophy, e.g. by the political scientist Jane Bennett.[78]

Vitalism is the holistic thesis that living beings, which we classify as organisms, form wholes that differ from the sum of their parts. The whole is the decisive carrier of the causality of the structure, i.e. we can explain the behavior of the parts of an organism better from the whole structure and thus from the functions of its parts for the whole than vice versa.

This thesis is by no means obsolete or even scientifically refuted, as is often claimed without evidence and historical confrontation with the great vitalistic approaches of Driesch or the famous vitalistic philosopher Henri Bergson, who was a contemporary of Driesch and a Nobel Prize winner for literature.[79] In contemporary physics, biology, computer science, theory of science and metaphysics, one speaks of "top-down causation" (the whole determines the behavior of the parts), which is sometimes located (e.g. by the famous cosmologist and mathematician George Francis Rayner Ellis) even in purely physical, as yet unliving systems.[80] This revives genuine vitalism, which in no way postulates that there is a mysterious life force; rather, it points out that living systems must be understood as wholes, in which parts and whole interact in a way that cannot be understood by analyzing the mechanics of the parts alone.

The controversy that some of his contemporary reductionist philosophers from the so-called Vienna Circle conducted with Driesch was by no means decided in favor of mechanism and reductionism. Today's life sciences are aware of the effects of our nutrition on the molecular level, which means that our genes are also determined by what we do as wholes.[81] There are, after all, effects of the whole on its parts, and vitalism has advocated nothing more. In any case, it would be a wild exaggeration to conclude from today's knowledge of the elementary, small-scale processes that take place in an animal that an animal is nothing other than the sum of its parts. That is simply not true.

The mechanism, which Descartes and after him La Mettrie almost took for granted, thus misses its object, the living. The application of the thesis of the animal as machine to the human being, which La Mettrie made, is interesting insofar as it logically follows the pattern that is at the heart of this part of the book: first, the human being projects the mechanistic concept of the animal onto alien forms of life in order to distinguish them from its own self-understanding; then it identifies itself with this distorted animality. The animal is seen as a machine, then we are conceived as animals, and thereby we also transform ourselves into machines. Mechanism was so successful because it nourishes the illusion of clarity and controllability of nature, not because it is true.

Animals Like Us? Korsgaard's Values

The influential Harvard moral philosopher Christine M. Korsgaard has recently presented a highly regarded approach to the justification of animal ethics which can be used to show how interconnected are our concepts of animal, human being and nature.[82] Korsgaard understands humans as those animals that are capable of science and ethics and therefore can adopt a perspective that is supposed to be valid for all animals. She thus tries to derive moral duties towards non-human animals, our *fellow creatures*, as she calls them, from our higher capacity for knowledge.

Thus her approach has an open flank that needs to be closed. This is due to the fact that Korsgaard does not go far enough with her insight that the human being is intertwined with the other animals. Her attempt to define the human being as an animal fails because – in a typical pattern with which we are already familiar – she does not give animals enough credit, seeing them as deficient beings, so to speak, as human beings who lack something essentially human (in her case, science and ethics). This is a variant of what the French philosopher Corine Pelluchon calls a "privative zoology."[83]

Privative zoology (*from Latin* privare = *to deprive*) *is a theory that understands non-human animals as animals in the same sense as we understand ourselves as animals but recognizes a deficiency in the other animals because they lack exactly that which distinguishes us from them (science, reason, morality, language, mathematics, or whatever).*

Korsgaard's starting point is the thesis that living beings – or, as she says, *creatures* – do not stand in any "absolute rank ordering of the importance."[84] Human beings are therefore not worth more than lions. But she immediately qualifies this a little: "Or rather, there is *almost* no place we can stand to make this sort of judgment[.]"[85]

That human beings are not worth more (or less) than the other creatures Korsgaard derives from the fact that any value that a living being can register (i.e. any meaning of an action or even of life as a whole) depends on the fact that the creature feels this value. Values are therefore subjective, as one commonly says, or, as she expresses it "tethered." The fact that I personally prefer to drink wine rather than beer is, in this sense, a tethered value. Because of my preferences, I give wine a value, a meaning in my life, which it does not need to have in the lives of other people or even other living beings.

Our environment is indeed shaped by the fact that, as living beings, we reshape it into an ecological niche. We do not simply find ourselves in a natural environment but change the place we find ourselves in on many levels – not least by affecting the atmosphere through our breath and (unfortunately) even more so through our cars, coal-fired power plants, etc. Korsgaard thus considers our valuations as a whole as projections, of which in reality nothing remains but the things in their mere materiality, to which we give value from the subjective perspective of our lives, value they don't have in themselves.

> [C]hairs and tables exist in the perspective of creatures who need or use furniture. Without such creatures, there might still be, say, wooden objects shaped in such-and-such a way, but they would not be furniture. If we were oval and swam through our atmosphere like fish, there would be no tables and chairs in our world. In a similar way, values exist in the perspective of a certain kind of creature, a creature who values things, in the sense of having evaluative or valenced attitudes towards things.[86]

Thus, Korsgaard believes that all life forms have their specific values, such that individuals of a species come to individual values in turn. But nature in itself, according to Korsgaard, has no value. Only living beings give it values from their own perspective, when they value nature. For Korsgaard, values come into the world through the fact that living beings value something. And this also provides her with a definition of animal:

> An animal – at least as I will use the term here – is a particular kind of living organism. An animal is an organism that functions, at least in part, by representing her environment to herself, through her senses, and then by acting in light of those representations. She is guided by her representations to get the things that are good for her and avoid the things that are bad for her, in the functional senses of good-for and bad-for. In order for an animal's representational system to do its work in this way, however, it has to have what I will call a "valenced" character.[87]

These considerations all sound as if Korsgaard consequently put the human being on the same level with all life forms. Thus, the human being would be an animal like all others, and that means for Korsgaard a living being which evaluates reality from its point of view and thus pursues its interests, first and foremost among which is its survival.

But Korsgaard would not be what she became famous for as an ethicist, namely a representative of a (somewhat) Kantian point of view, if she did not somehow have room for the thought that our ability to think and act morally is central and goes beyond the merely relative valuations of other creatures. For Korsgaard believes that only humans are able to form "science and ethics"[88] because only we are able "to grasp a terrible fact about the natural world,"[89] namely that

> … *it* exists completely independently of our own interests; that it is the work of mechanical forces that operate with no regard for us or for the other animals; that it is a world in which there is no guarantee that things will turn out well for ourselves or for other human beings or for the other animals or for life itself.[90]

Now Korsgaard wants to derive ethics in general and animal ethics in particular from this modern nihilistic point of view, which again is interesting because it shows how our pictures of the world, the human being and the animal are interwoven in our representation of humanity.

Modern nihilism is the thesis that nature in itself has no value, meaning or significance. It does not care about us and our needs but is simply a material-energetic total structure, in which living beings have at some point emerged. These beings evaluate their environment, which is the origin of values. In itself, nature is entirely devoid of value.

Korsgaard makes this very nihilism the basis of ethics insofar as she ties it to our ability to refrain from our egoistic interests and to recognize that what we value as individuals or as a pack has no objective value in itself. Thus, there is room for the recognition of the values of other people and living beings as well as for the realization that they are not "into the world for *our* use."[91]

Drawing on animal-philosophical considerations, Korsgaard assumes – analogously to Kant's categorical imperative – that only living beings who can ask themselves what would be right from the point of view of all concerned are capable of answering ethical questions.

Ethical questions are concerned – and in this Korsgaard agrees – with what we should do or refrain from doing par excellence, *i.e. simply because we are human beings. The ability to formulate and answer ethical questions presupposes that one can refrain from one's own interests in favor of others.*

At this point, Korsgaard launches another maneuver that is familiar from the Kantian tradition. I mean the following consideration, which can be understood even without indulging in the details of Kant scholarship: insofar as we value anything because it suits our interests, we always share a formal property with all other living beings who also care about something – their offspring, food sources, sexuality, or whatever. All who care about something are selves, i.e. they locate themselves in a structure that consists of them as well as of something that is not themselves. To be someone, i.e. oneself, presupposes that one experiences one's life on the one hand as a self-fulfillment and on the other hand as a foreign event. Or even more briefly: to be someone means to be able to distinguish between self and non-self. Using traditional philosophical language, Korsgaard calls this capacity self-consciousness.[92]

But this is far from answering the question why someone can actually distinguish between good and evil and thus between something that everyone should do or everyone should refrain from doing. Therefore, it requires the reflexive additional component that living beings, which can comprehend themselves as self-conscious sources of value, are ends in themselves. In Korsgaard's reading of Kant, this is justified in such a way that we as human beings have dignity and thus an unconditional value that cannot be calculated against anything, for we can conceive of ourselves as a source of value without which there would be no absolute, unconditionally valid values such as human dignity.

According to Korsgaard, through our structure as self-conscious sources of value we bring forth the absolute value of being able to value. Therewith, thanks to our existence, we pull ethics (like the Baron Munchhausen) by our own hair out of the swamp of mere animality. Korsgaard thinks that, from the fact that living beings set relative values (by being hungry and valuing cheese, for example), it follows that the source of value itself has an unconditional and not merely relative value. Yet, this discovery presupposes self-consciousness (science and ethics) which make us special sources of absolute value.

Unfortunately, this broadly Kantian maneuver is full of holes in many places. Above all, it leads to two consequences that are difficult to sustain and whose designation will take us further.

First, Korsgaard's construction runs towards a strong anti-realism, which she formulates and best illustrates in the following passage:

Strong anti-realism according to Korsgaard: *"Moral standards are nothing more than the standards of behavior that we can all want people to adhere to. But contrary to what you may have thought, it is not that everyone can agree to this behavior because it is morally right. Rather, it is morally right because everyone can agree with it."*[93]

Value realism would be the view that there are objective values knowable by us that exist independently of our recognition of them, while *value anti-realism* claims that values exist only because we recognize them.

Anti-realism has some weaknesses, mainly because it cannot explain how one can evaluate and justify concrete actions and decisions within its framework. Consider the many ethical problems that played an important role in our daily lives during the coronavirus pandemic. Can everyone agree to mandatory masking in elementary schools at a given incidence level, a curfew at 10 p.m., or even a triage procedure in the event of a shortage of medical supplies? The ethically complex situation in a concrete case of action can hardly be solved by reaching a consensus in which everyone participates. Therefore, the view that in specific cases there is something to which all can agree is an over-extension of the idea that moral values are constitutive, i.e. conceptually bound to consensus. The idea of morality as a kind of agreement exaggerates the correct and important observation that there are many cases in which we agree ethically and recognize the same moral facts as self-evident (e.g. that we should not kill anyone). Nevertheless, it does not follow from the fact that ethical consensus is possible that ethics merges into consensus, that there are therefore no objective values at stake in ethical discussion.

But regardless of whether ethics can be grounded in the construction of ideal consensuses (which never really exist in difficult questions!), Korsgaard's version of a strong anti-realism leads to an unjustifiably anthropocentric position. For, according to Korsgaard's position, why should we not radically reduce biodiversity (for example) if this would benefit humans? If we had to exterminate some other animal species in order to live healthier, longer and happier lives, this would not be morally justified in virtue of all humans agreeing to it. While we would prevent some human deaths by killing all bees and wasps, it does not follow that it would be morally right to do so if all people were in favor of protecting the allergy sufferers among us (of which I am one). Although bees and wasps are particularly dangerous to me (due to an allergy) and to many other people, I would consider it a moral mistake, indeed an outright evil idea, to exterminate bees and wasps in order to protect human lives. And even if everyone – allergy sufferers and non-allergy sufferers – agreed that bees and wasps should be exterminated to protect people, it would still be a moral mistake to carry out such genocide on these insects.

Of course, Korsgaard has an answer to this in that she assumes that we must take into account the point of view of other living beings (fellow creatures) when we ask ourselves what is in the interest of all. But animals – not to mention our shared environment – cannot be considered ethically independent of us as an absolute source of value for Korsgaard, so they will not be invited to have a say in the "parliament of things," as the postmodern French sociologist Bruno Latour puts it.[94] Our fellow creatures can only be objects of ethics for us to the extent to which we agree to recognize their relative values as compatible with the absolute value of human ethical valuing.

This is a result of Korsgaard's anthropocentric starting point. By seeing humans as the crown of creation, distinguished from all other animals by their higher insight and by their production of universally shareable, moral values by consensus alongside crude survival interests, the other living creatures become "fellow creatures."

This is a degradation insofar as the other animals are once again seen as deficient beings from the human point of view, i.e. as human animals without the specific humanity of science and ethics – a clear case of privative zoology (see p. 68). Thus, the question does not even arise to what extent we humans can perhaps learn something from other, similarly prosocial creatures, even in moral questions. Rather, the animal kingdom appears as a pure state of nature, in which only hard-core, relative interests apply, without the actors ever being capable of real moral insight.

But how can Korsgaard be so sure that none of the other living beings are able to recognize good and evil? It is true that we humans discuss in our languages what we should do or refrain from doing, and can therefore communicate and formulate our reasons for doing things differently than what up to now we know of other life forms (who knows what whales, bees or forests discuss with each other?). But it does not necessarily follow from the "silence of animals," as the French philosopher of animals Élisabeth de Fontenay calls it, that they are unable to make moral judgments.[95]

Alice Crary – Inside Ethics

This is precisely where we can make progress with the help of what I consider to be a groundbreaking approach, which goes back to Alice Crary. In her book *Inside Ethics*, she shows that ethical thinking cannot, in principle, succeed "*from outside.*"[96] By this she means that the question of what we should do for moral reasons in a specific situation of action, which we always share with other humans and non-human living beings, can never be approached only by looking at reality from the largely viewpoint-free standpoint of an empirical observation.

If you see a dog howling in pain because someone stepped on its paw, you don't need a study to determine that you have inflicted an evil on the dog that should be avoided. If you step on a dog's paw for fun, you are committing a moral wrong and doing something that falls within the realm of evil. And just as one should not rob human parents of their child in order to eat it, it is probably the morally right thing not to order the Wienerschnitzel, because even calves have parents who care that their child does not end up on the plate of another species.

This argumentation presupposes, of course, that the unimaginable, terrible suffering that is generally inflicted on human parents and guardians when they are separated from their children (as a father, I do not want to imagine at this point what evils are inflicted on children and those who are responsible for them on a daily basis!) also occurs in an analogous form in other living beings. As I learned on a recent visit to a farm, bulls often do not distinguish between their own offspring and the offspring of other members of their species in their sexual desire and aggressive behavior, which is one of the justifications for separating calves from their parents. This of course does not justify the consumption of calves, but is only an indication that our human kinship relations and social categories are incomparably more complex and especially ethically more elaborate than those of many other living beings with whom we easily identify on an intuitive level. We must recognize that humans are radically different from other living beings in many ways, which is why the concept of animals is traditionally also a concept for life forms that are deeply alien to us – which is why many cultures and religions have deified them, so to speak, as a precaution.

We experience reality as morally charged, thus in such a way that we immediately feel that we should do or refrain from doing certain things. We have no reason to believe that other living beings, which also judge normatively, dividing their environment into pleasurable and non-pleasurable circumstances, are outside ethics, and are as it were pure natural occurrences that have no intrinsic value. On the contrary, everything we know about life today speaks for the fact that other living beings also experience values and so have value concepts (by preferring certain landscapes to others for ethical reasons, for example), which they nonetheless do not put down in writing or transmit in any other way or change with a view to moral progress. This means that we ought to expect that we could learn something from other modes of valuing such that the moral perspectives of other life forms can precisely not be reduced to relative values. In sum, humans might not be the only sources of absolute value.

No matter how it may feel to be a dog, a bee, an octopus or a bat, everything points to the fact that reality is experienced by these creatures not simply as an anonymous event but as something that has intrinsic value and not just value relative to their interest. The clearly prosocial behavior which we find outside of the human context (and already on the microbiological level) can be classified as an indication of value cognition. It does not follow that bees have an explicit ethics and even build a state on it, but that they are able to experience the value of reality. Just as bees and other creatures certainly recognize whether there are many or few flowers in a meadow without being able to develop a mathematical theory of numbers or sizes, they might be capable of ethical value judgment in ways currently largely inaccessible to most of us.

Crary rightly rejects modern nihilism, i.e. "an enormously widely accepted metaphysical worldview on which the real fabric of the world is in itself bereft of moral value."[97] According to modern nihilism, insofar as human beings and other living beings belong as bodies to a value-free layer of reality, we are thus largely – and the other living beings almost entirely – outside ethics. This means a justification is needed as to why we should behave with moral rightness at all. According to Korsgaard, this is given by our reflexive ability to see ourselves as a source of value, from which it immediately follows that only our rational, reflexive and self-conscious judgment is a real source of value. Thus, in Korsgaard's view, the ancient original division persists of the human being into an animal and a rational side, which then subsequently ensures that the other living beings are regarded as mere animals yet again.

Anthropological dualism is the thesis that a human being consists of two parts: an unreasonable animal part and a specifically human part, which is mostly regarded as superior. The main problem with this thesis appears when the unreasonable part is misunderstood as value-free or even as inferior, as a mere 'animal in us.'

Crary avoids this anthropological dualism from the outset by bidding farewell to the idea that reality is intrinsically value-free. This allows her to bring the lived experience of non-human animals into view and to accord them full membership in the space of those beings who are capable of experiencing value without having to produce or project it into an otherwise meaningless universe. As I will argue in the second part of this book, this conception of an intrinsically value-free or at best cruel nature, which doesn't give a damn about all of us, turns out on closer examination to be a value judgment. This is precisely one of the reasons why science has not discovered that the universe is meaningless.

In Crary's view, modern nihilism is replaced by a conception of reality according to which we do not have to defend our ethical attitudes against nihilism by means of complicated philosophical arguments, because we always exist "inside ethics," i.e. inside value judgments which include other living beings since time immemorial:

The world is filled with things that are endowed with objective moral values, human beings and animals. Hence we are right to take at face value a natural understanding of moral judgments as world-guided assessments of appropriate responsiveness to these values.[98]

Crary intervenes in the right place with her approach. When it comes to questions of ethics in general, and animal ethics in particular, the relationship between humans, animals, nature and reality is under scrutiny. One of the reasons we are on the path of an obviously unsustainable self-extinction driven by the accelerated engine of modern industrial capitalism is that we have a false conception of ourselves that sees human beings primarily as spectators – a part of a reality that is decoupled from nature. This blinds us to the myriad interactions that take place between our activities and the so-called environment, which does not surround us but rather permeates us. The environment is not around us but within us (think of the many organisms that populate you while you read this sentence, or the sheer fact that all of us are parts of physical fields and do not somehow observe them from the outside). Thus we are imperceptibly interfering with nature. Every click, every Google search, every post, every like, every message, every video conference in which we articulate our fears and anger against those who drive climate change generates and uses data that would not exist without gigantic server landscapes devouring enormous amounts of energy. In this sense, unfortunately, our digital life-world today actually resembles the Matrix as depicted in the most recent installment *The Matrix Resurrections*. People have literally nested themselves in the machine room of a dystopian future, where their mental lives are largely digital while they try to cultivate one little plant or another in order to compensate for their self-inflicted harm in which all of us participate, including those who think they are innocently fighting the destructive forces of modern industrial capitalism. It will take more to overcome the dark side of modernity than an attitude of critique and complaint from the progressive margins. What we need first and foremost is a novel conception of ourselves and our position in nature, one that overcomes the wrongheaded assumption that we are thrown into a meaningless universe which we happen to inhabit alongside other animals on an otherwise unimportant planet.

Subjectivity and Objectivity – Why We Aren't Strangers in Nature

According to a widespread idea, *science* is the paradigm of an objective determination of facts to which we must orient our social conditions. In this context, many believe that conspiracy theories and fake news must be countered by science in order to protect democracy. However, there is no such thing as a uniform, all-encompassing form of objective knowledge or even a single method called science; rather, there are very different disciplines that gain knowledge with different methods. To put it in the words of the founder of ecology (to whom we owe the concept of the environment), Jakob Johann von Uexküll:

> Nowadays the word "science" is used in a ridiculous fetishism. Therefore, it is probably appropriate to point out that science is nothing else than the sum of the opinions of the researchers living today.[99]

It does not follow, of course, that there are no established realities or facts which (thanks to scientific methods) we cannot reasonably dispute at a given time. Of course, this also applies to the complex crisis phenomena with which we have to wrestle today. For instance, it is objectively ascertainable that there is human-made climate change that is connected with our scientific-technological civilization, and which has catapulted us into a fossil fuel age with gigantic energy consumption. We know this thanks to various scientific disciplines, without which human-made climate change would not exist on today's catastrophic scale. Without modern physics, chemistry and geology there would be no fossil fuels, engines, or nuclear and biological weapons. To believe that science and technology alone can solve the problems for which they are largely responsible is a naïve mistake from which we urgently need to free ourselves. Paradoxically, the scientific disciplines thanks to which we can determine the existential risks we are facing are at the same time responsible for the very existence of those existential threats.

Many scientific disciplines (from applied physics to biology) are researching the effects and consequences of climate change using natural-scientific methods, and they can and, indeed, must be used to try to develop more sustainable forms of surplus value production. But without sociological, historical, political and cultural scientific research and methods, we cannot sufficiently cover the climate issue. For example, India or China must be studied as complex historical and social entities if we want to understand which form of sustainability is compatible with which social formation. It is not enough to rely on the fact (which cannot be meaningfully disputed) that we must reduce humanity's CO_2 emissions as quickly as possible. At the same time, historical, cultural, social and even religious dimensions of human life must be taken into account, and these cannot be meaningfully explored by natural science. In short: there is a plurality of sciences which at best have in common that they claim knowledge by means of historically developed, objective methods and thereby achieve knowledge. But there is no single unified knowledge formation called "science." Moreover, it is not only the sciences that will cope with the climate crisis, because new practices and socially organized forms of life and experience are also needed that cannot be produced scientifically. Why, for example, should we import wines from all points of the compass when good wines are now produced everywhere nearby? We should learn again to appreciate regional products instead of always wanting to access global goods. These are not scientific questions but questions of aesthetics, desire, advertising, and so on.

The demand that we should get behind science, which seems attractive as an anchor of orientation in view of the complex crisis situation in which we find ourselves, overlooks the fact that those problems which we would now like to solve with the economy, science and technology have arisen because moral, human progress was decoupled from scientific-technological progress only a few decades after the French Revolution.[100] In this development, the epistemologically and scientifically untenable idea prevailed that objectivity consists in describing reality in the form of mathematical models in order to make it predictable and controllable within this framework. But it is exactly this instrumental understanding of nature that has allowed the modern combination of natural science and technology to generate products that are devastating our planet. Far from redeeming humanity and explaining to us how to live a proper life overall, the much invoked science in the eminent singular has made the modern energy and consumer society possible, which we now want somehow to bring back under control with even more science.

Behind the idea that we simply need more science and technology to solve our crises lies a whole series of errors, first and foremost an error about objectivity. The instrumental understanding of nature regards as objective primarily that which is without our subjective experience, i.e. without our individual perspective.

---

*This view of objectivity can be described as **metaphysical scientism**. It claims that objectivity consists essentially in the fact that we strip off our subjectivity in order to develop a worldview which grasps reality as it is in itself. This worldview presupposes empirical studies, measurement data and mathematical models as its proper means of articulation and communication.*

Metaphysical scientism has become a seminal notion in the course of modernity, as the historians of science Lorraine Daston and Peter Galison have shown in their book *Objectivity*.[101] It is not a matter of course but a one-sided view of our human capacity for knowledge.

Our cognitive ability consists not in stripping away our subjectivity (and thus our value judgments) in order to recognize an in-itself value-free reality but, rather, in recognizing how reality is from the standpoint of a human being. Thereby our point of view does not falsify reality; rather, it belongs integrally to it. We can recognize reality only because we are a part of it.

In this regard, it is often said that twentieth-century quantum physics has upset the false picture of objectivity criticized here by discovering the role of the observer or, less misleadingly, the not negligible contribution of our intervention in every physical system through which we can obtain physical knowledge in the first place. According to this opinion, quantum physics should have already physically grasped the human point of view more than one hundred years ago. But this is not correct, or at least it results only from some interpretations of quantum physics, but not from its impressive and precise mathematical findings. Quantum physics came into being by pioneering mathematical-physical achievements, which have influenced our conception of the functioning of the universe independent of us. But it is not able to capture the subjective point of view or consciousness, which is not really its topic. What quantum mechanics entails for metaphysics and epistemology is a hotly debated topic. Yet, it is misleading to believe that it tells us something about our subjective point of view which certainly cannot be found anywhere in its foundations or equations.

Precisely because physics has the task of investigating a reality which is not colored by our experience but is structured by mathematical parameters which actually do not care about our individual interests and points of view, it could foster the erroneous view in modern times that we, as experiencing, mental, living beings, do not really belong to reality. Because thoughts cannot be measured by physics, what takes their place is the measurement of brain states and thus the erroneous belief that thoughts are ultimately identical with brain states. The human being became increasingly a stranger in the universe by the achievements of modern natural sciences.

As we saw, Crary rightly opposes this with the apt reference to the fact that we experience reality as value-laden, an experience which does not (and need not) fit into our physics. In this way, she comes close to the philosophical school of thought known as phenomenology, which was decisively developed by the great mathematician and philosopher Edmund Husserl and is still one of the most influential innovations in twentieth-century philosophy. In simplified terms, one can formulate its starting point thus:

Phenomenology *investigates the way reality appears to us. In doing so, it does not try to abstract from our way of experiencing reality, from our subjectivity, but conversely assumes that it makes no sense to ask for a reality independent of humanity, a reality in which we nevertheless show up.*

The realities in which we live are always shaped by our copresence.[102] Therefore subjectivity and objectivity are not mutually exclusive but interact in our human perspective. Subjectivity and objectivity are intertwined in the mind: the reality we experience always includes us.

This it not to say that everything that is real somehow depends on our perspective or experience. It means only that reality, as we know it, essentially involves us. We cannot meaningfully conceive of reality as a whole in such a way that we play no role in our understanding of it. We are as essential for reality's unfolding as everything else, and we are even central to any understanding of how we come to understand reality's existence and structures in the first place.

Precisely because of our central epistemological position, we can dissect many facts and analytically determine that they would be in every relevant respect as we find them, even if we had never turned to them. The fact that there are many millions of galaxies, for example, is not a product of our astronomy but, rather, something that we have discovered scientifically. There are of course infinitely many facts which are completely independent of us in an undemanding and somewhat obvious sense. But from this it does not follow at all, contrary to the claims of metaphysical scientism, that it belongs to the definition of objectivity and reality that they are independent of us.

Against this background, a New Realism has developed in contemporary philosophy, whose central ideas can be summarized for our context as follows.

New Realism assumes that our experience of reality, the mind or subjectivity, belongs to reality on an equal footing with that which we justifiably classify as independent of us. Therefore objectivity does not paradigmatically consist in the fact that we abstract from subjectivity.[103] *The objective includes the subjective (and vice versa): objectivity and subjectivity are mutually contained within each other.*

Full recognition of this fact opens the way to a *new moral realism*, i.e. to the insight that the reality of our experience of value, and thus the basis of our practice of moral judgment, is not an illusion but, rather, a form of cognition that allows us to address those normative guidelines that humanity urgently needs without succumbing to the erroneous idea that we merely need to consult *science* regarding the hard facts (such as human-made climate change) so that we as a society as well as individuals know what we should do. The hard facts of climate change, which we express in relevant data and play out in unfortunately quite realistic future scenarios, teach us nothing as such with regard to the question of what moral attitude (and thus what understanding of humanity, nature and life) is necessary to change our individual and collective lifestyles. Sustainability does not result merely from technological solutions but has a lot to do with how we want to live in the first place, and thus with the crucial question which life we consider meaningful.

What the natural sciences have discovered within the framework of the impressive methods they have developed in recent centuries, long since beyond meaningful dispute, does not immediately have normative consequences, does not by itself dictate to us what we should do. For the object of the natural sciences is first of all value-free: the natural sciences explore the structure of non-moral facts.

Therefore, to bring Crary back into the conversation, it matters that "we 'widen' our view of the objective realm so that it encompasses morally significant things."[104] Such a widening does not undermine anything we know about the non-moral facts but integrates such factual knowledge into our ethical practices.

It is nonsense to demand that we *"unite behind the science,"* because this is also done by technocratic regimes, military dictatorships and the automobile industry, which is now considered by many climate activists to be evil *par excellence*. The natural sciences have clearly shown that there is a very dangerous human-made contribution to climate change, which holds great dangers and injustices. But it does not follow from this how and whether we should avert climate change at all. The (correct!) value judgment that we must radically rethink our social and economic systems in order to strive for a new way of life that allows humanity to live a dignified life on our planet in coexistence with other living beings does not follow from the scientifically known facts. These have been known for decades without us so far having changed our lives appropriately in order to lead a more sustainable, calmer and ultimately more spiritual life in which self-knowledge and the pursuit of wisdom would be more important than short-term consumption of goods eventually doomed to be scrap.

Values are a separate part of reality: they form irreducible fields of sense, i.e. they cannot be replaced by any value-free field of sense. The fields of sense of values can be grasped by living beings who experience the universe not only in colors, forms, smells, sounds, sonar impressions, etc., but also in terms of values, by experiencing what happens and is done as good, evil or neutral. Thereby values are just as little illusions as colors and smells, even if they can only be grasped if we take our own presence in reality into account.

The New Enlightenment in the Age of Living Beings

Given the complex crisis situation in which human beings find themselves today, calls for a New Enlightenment have been sounding for some time. Corine Pelluchon emphasizes:

> This must have a positive content and present an emancipatory project on the basis of an anthropology and an ontology that takes into account the challenges of the twenty-first century – and these challenges are at the same time political and ecological in nature, as well as linked to our coexistence with other living beings, both human and non-human.[105]

New Enlightenment, as I propose to conceive of it, is based on three assumptions: moral realism, universalism and humanism.[106]

Moral realism is the assumption that there are objectively correct beliefs about what we should or should not do merely insofar as we are human.

Universalism is the idea that what we should do or refrain from doing for moral reasons is valid for all people, i.e. it is cross-cultural and in principle also recognizable for all people in their respective contexts.

Humanism states that ethics, as reflection on what we should do or refrain from doing, is anthropogenic, i.e. it arises from the human form of life, without therefore being anthropocentric, i.e. directed only at humanity.

In view of the challenges of our time, the New Enlightenment is convinced that we must look for ethical solutions to complex problems, such as the ones associated with the climate crisis.

What we find out about nature through scientific-technological research is correct; it is based on the facts and thus discovers how nature behaves under conditions established in the form of experiment and theoretical construction. Thanks to this attitude towards nature, we have known for decades that our one-sided consumption- and growth-oriented conception of successful economic activity exceeds planetary limits and threatens humanity with its own self-extinction.[107]

Which solution a certain partial differential equation has, how fast we reach which tipping point at which levels of CO_2 emission, or how nerve cells are structured are all facts. But they can be normatively categorized in different ways – ethically, politically, socially, economically, religiously, aesthetically, and so on.

Therefore, the New Enlightenment rejects the reductionist belief that we can overcome the complex crisis situation of the twenty-first century by "uniting behind science" or even by marching for science. It is certainly dangerous that there are many people who deny factual scientific knowledge, and there are many reasons for this. (In addition to fake news, social networks and other symptoms of decay in our media, the reasons include a variety of complex socio-cultural and psychological factors.) But these causes for the widespread syndrome of denialism (i.e. the denial of secured, factual knowledge) cannot be solved by sole recourse to science and technology.

My criticism of the one-sidedness of the concept of science, which is inherent in the English term *science*, is by no means connected with the erroneous idea that scientific findings are false or irrelevant for normative questions. However, they can be normatively evaluated and used in different ways, as can be learned from the ethical discussion about nuclear power or the recent discussions in the novel field of ethics of AI. Insofar as scientific-technological progress is decoupled from normative questions, there is no point in uniting behind the fiction of a singular science. To quote Pelluchon again, who rightly takes up the classic critique of Enlightenment by Max Horkheimer and Theodor W. Adorno at this point:[108]

> It would be naïve to believe the development of science was a sufficient condition for global progress, which would then also manifest itself in the moral and political spheres. The Enlightenment did not foresee that objectifying thinking carried the risk of a drift of reason. This manifested itself in the instrumentalization of science, which had degenerated into techno-science, by the economy, and in a dehumanization as a result of a loss of contact with the living world.[109]

The way of thinking about complexity Pelluchon proposes is essentially ecological in that it entails the integration of our life processes into an environment through which we are vulnerable, mortal, but also able to use these facts about ourselves as animals for moral progress.

There are moral facts which consist in the fact that answers to ethically relevant questions can be true or false. For example, the answer to the question, "Should I save the toddler who fell into the pool in front of me from drowning?" is clearly "Yes." It is a moral fact. Other examples of moral facts include all the war crimes taking place as you read these lines, such as the shelling of hospitals. Such actions are clearly evil.

So there really is something we should do or refrain from doing (good and evil), as well as a gray area in which we look for orientation. We are far from having addressed all the relevant ethical questions of our age. One reason we find ourselves in a crisis situation lies precisely in the fact that we do not treat normative questions as such. Instead, we repeat the same mistake that got us into the modern predicament, only at a higher level. The foundations of our impending self-extinction were created in the first place by decoupling ethical (and thereby also philosophical) thinking about ourselves as human beings from techno-scientific progress.

More science and technology alone will not solve the problems but at best shift them, as digitalization shows. On the one hand, digitalization makes a (rather superficial) contribution to CO_2 reduction. Thanks to video conferencing, we fly less, we print less, and, thanks to modern high-performance computers, we gather knowledge about the climatological relationships we must understand as precisely as possible in order to protect ourselves from impending catastrophes. But, at the same time, the data that has to be processed and stored on ever-larger servers in turn produces dangerous junk and CO_2. As if that were not enough, without the progress of modern computer science, which is welcome in itself, there would not even be those systems that produce fake news and other forms of disinformation – social media first and foremost. In general, given the modern knowledge economy, there is paradoxically an overproduction of knowledge. So it is not even possible to survey scientific factual knowledge as a whole or to assume that there is such a thing as science in the singular, a science we'd only have to listen to well enough in order to tackle our normative problems.

Scientism is the false notion that there is a singular thing called science, which will ultimately someday be omniscient. Scientism believes science will prevail socio-economically through experimentation, rational theory construction and expertise, and that it will finally rid humanity of all ills – such as disease, death and political uncertainty.

The New Enlightenment is directed against scientism in the name of moral and thus human progress, without therefore making the mistake of denying scientific knowledge. On the contrary, the project of a New Enlightenment is based on massive transdisciplinary and trans-sectoral cooperation: the various sciences (which, of course, include the humanities and social sciences in the same rank and sense as the natural and technical sciences) must work together with the other subsystems of society – such as the economy, politics and civil society – to develop an image of the human being that, as much as possible, aligns with the body of knowledge of the twenty-first century.

This also involves changing our structure of needs and our view of the relationship between humans, nature and animality, which Pelluchon calls the "schema" (*le schème*).[110] She defines the schema as a matrix of modes of production and reproduction of our human form of life that is socio-economically orchestrated. According to her, the weakness of our present-day scientistic matrix consists in the fact that we search for exclusively technocratic solutions to our crises, without researching their spiritual, existential, psychosocial grounds and changing these through novel practices. Our relationship to ourselves and to others is disturbed because we have the wrong socially shared idea of a purpose – or, more precisely, basically no idea of such a purpose. And that means no vision of the good that lives up to the hyper-complex situation of the twenty-first century.

Kant's Four Questions – Being Human is an Answer to a Question

The human being is a living being who determines itself. This self-determination basically consists in the fact that we as individuals (but also as collectives) lead our lives in the light of a conception of who we are and who we want to be. Our animality is part of this self-determination because we are not simply animals and also something else (rational, linguistically gifted, etc.). It is constitutive for us as human beings that we ask ourselves what the meaning of our lives actually is, in order then to answer this question through our concrete individual decisions as well as through the establishment of social relations. How people live at any point in their historical self-determination thus depends on how they see themselves. In this sense, people are answers to the question of who or what the human being is.

Our self-determination thereby takes place in the space of facts and not only on the basis of feelings. We are by no means living in a post-truth age, which Angela Merkel once defined in 2016 as follows: "This probably means that people are no longer interested in facts; they follow their feelings alone."[111] Our self-determination is therefore prone to error, which means we can grasp truths about ourselves or also miss them. Self-determination is therefore not an infallible project. Moreover, human beings are essentially social: we correct each other by mutually observing and informing each other in order to maintain a society in which there are diverse projects of self-determination, which are nevertheless by no means all right and proper.

A simple but weighty example illustrates this thought. We either have an immortal soul or we do not. Regardless of the facts, some people believe they have one, while others doubt it. Thus it is clear some members of the human race are mistaken about themselves – although unfortunately this does not yet tell us who is right. Because our images of humanity are prone to error, it is crucial we all work together on our human self-image. This requires recognizing that there is a wide spectrum of human self-determination and that no one has it entirely in view.

The New Enlightenment is a cosmopolitan project. Thus, it cannot function without transcultural understanding. Its universalistic claim can only be redeemed if we ask how different traditions, different collectives (cultures) and individuals understand themselves as human beings. This presupposes we explore and reconcile cultural differences in our pictures of the human being, nature and animals in order thereby to give the most complex answer possible to the central question of philosophy: What is the human being?

At this point, it is essential to remember Immanuel Kant once again, who not only made a significant (by no means only positive) contribution to the Enlightenment of the eighteenth century, which many today virtually equate with the Enlightenment.[112] Kant, like most philosophers before and after him who have had an impact on society, not only published academic writings in which he developed a theoretical and practical philosophy of reason. He was also socially engaged and addressed the general audience, the public. Insofar as philosophy has the task of putting its results and questions to the test outside of specialist circles, Kant speaks of the "world concept" of philosophy, which he considers to be "cosmopolitan" because it is about humanity as such, i.e. about all human beings, who – contrary to what Kant himself unfortunately thought owing at times to massive racist prejudices and aberrations – cannot be divided into closed cultures or even "races."[113] In a remarkable passage found outside his famous major works, in a logic lecture, he states:

The field of philosophy in this cosmopolitan meaning can be brought down to the following questions:

1) What can I know?
2) What should I do?
3) What can I hope for?
4) What is the human being?

Metaphysics answers the first question, *morals* the second, *religion* the third, and *anthropology* the fourth. Fundamentally, however, we could reckon all of this as anthropology, because the first three questions relate to the last one.[114]

Which cognitive and other abilities we ascribe to ourselves as individuals or as human beings depends on which human self-portrait we prefer. The image of the human being therefore determines the further development of our abilities, which are not predetermined once and for all. We do not know everything the human being is capable of. Of course, Kant was not the first to come up with the idea that the human being, thanks to its capacity for self-determination, is an open, free living being capable of many things, especially the destruction of nature. In the famous verses of the chorus in Sophocles' *Antigone*, the chorus already states:

Many the wonders but nothing is stranger than man.
This thing crosses the sea in the winter's storm,
making his path through the roaring waves.
And she, the greatest of gods, the Earth –
ageless she is, and unwearied – he wears her away
as the ploughs go up and down from year to year
and his mules turn up the soil.

Lighthearted nations of birds he snares and leads,
wild beast tribes and the salty brood of the sea,
with the twisted mesh of his nets, this clever man.
He controls with craft the beasts of the open air,
walkers on hills. The horse with his shaggy mane
he holds and harnesses, yoked about the neck,
and the strong bull of the mountain.

Language, and thought like the wind
and the feelings that govern a city,
he has taught himself, and shelter against the cold,
refuge from rain. He can always help himself.
He faces no future helpless. There's only death
that he cannot find an escape from. He has contrived
refuge from illnesses once beyond all cure.
Clever beyond all dreams
the inventive craft that he has
which may drive him one time to good or another to evil.[115]

We humans are capable of good and evil. This makes us dangerous, to ourselves and to other living beings. Because we live together with the other living beings and because we are the paradigmatic animal, we are obligated to think through animal ethics. If we were not animals but disembodied minds or angels (theologians are not unanimous on whether angels have bodies …), we would also not need animal ethics and an ethics of nature.

Thus it is important to explore the human self-relationship with regard to our own animality, which is part of our picture of the human being. The protection of animals and the environment cannot be achieved by putting our own needs aside, because we are also animals (and, if am right, we are actually the only animals in an important sense which makes us responsible for other living beings).

*The type of **ecocentrism** popular among environmental thinkers and activities, according to which the environment (whatever that is supposed to be) is the main reason for urgent human action, even for a complete transformation of our societies, is a mistaken view. For neither the environment nor nature alone can be a given for ethics. What we owe to nature cannot be derived from nature alone; we need to take ourselves into account for any determination of our environmental rights and duties.*

Therefore, there is no need for decentering or even overcoming humans in the sense of today's widespread posthumanism and transhumanism in order ethically to justify animal welfare and environmental protection. Ethics is not conceivable without human beings.

Posthumanism *assumes that any distinction between the human being and nature is ultimately obsolete and is due to outdated ways of thinking that set the human being apart from its environment and thus directly or indirectly see it as the crown of creation.*

Such a privileged positioning is replaced by the concept of *Gaia*, prominently advocated by the French sociologist Bruno Latour, for example, according to which we should understand nature as a complex network of actors that together form a living network in which we are, at best, nodes but not a central event.[116]

Transhumanism, on the other hand, is associated with the idea that we should transcend humans in favor of a higher, possibly technologically produced life form. As if we were about to witness the coming into being of the next stage of evolution (possible through our own making, in the form of superhuman AI).

This idea is often associated with science-fiction fantasies of a coming superintelligence, or at least an artificial intelligence we will produce, only to be replaced by it as the preeminent being on the planet.

It is precisely against such anti-enlightenment projects of human self-abolition that the humanism of the New Enlightenment is directed. We grasp the idea of human dignity and recognize it as a guideline for our actions because the human being is something special, a specifically minded living being which can become aware of its responsibility towards its fellow human and other living beings. Of course, our special capacity for insight does not entitle us in any way to reduce biodiversity or even to devastate the entire planet. On the contrary, every acceptable humanism commits itself to the main idea of modern environmental ethics, which was formulated above all by Hans Jonas, as the pioneer of an ethics of the living, in his *The Imperative of Responsibility*.[117]

Because we are capable of moral insight as well as technological superiority with respect to other living beings, we have moral obligations towards them – an idea which of course is not specifically 'modern' (think of animal ethics in Hinduism or Buddhism) but, rather, deeply human.

The Human Being as the Animal Who Doesn't Want to be One

As should have become clear by now, that the human being is an animal is not an obvious fact we can establish today by simple recourse to evolutionary biology. For the thesis is associated with a variety of theories and ideas, and some of these must be rejected. In particular it is not correct to say that the human being is *merely* an animal.

Specifically, this means a human being is neither identical with nor absorbed by the fact that a certain animal body or a certain organism is attached to it. The living being I see in the mirror is not my whole self but a part of me – unquestionably a quite important one. What I am (and in this I do not differ from you) is a currently embodied thinker of thoughts who takes a certain point of view. This point of view naturally includes feelings and experiences of different kinds. Thinkers of thoughts are not emotionally cold Mr Spocks or purely rational thinking machines but essentially living beings. However, our thinking and our life are not connected in our physically or medically measurable body but in what is traditionally called "mind" [*Geist*].

*In general, we can understand **mind** as the capacity to lead one's life in light of a conception of who or what we are. In this context, it is irrelevant for the existence of the faculty of the mind whether it conceives itself correctly, i.e. as mind, or falsely identifies itself with something that we are not.*[118]

Mind should be described as a whole. This whole has parts, which include our conceptual or intellectual faculties. As minded beings we are intelligent, attentive, conscious, reasonable, etc., and have many other faculties that allow us to recognize, understand and explain the realities in which we live. From the point of view of the mind, we arrange, evaluate and cultivate our individual abilities, and this is that of which our culture consists. In this way, our abilities play a role in our self-image, which is essentially shaped by the fact that the human mind is social.

Mind in the outlined sense is, mind you, not a specific difference that distinguishes humans from 'animals,' because this comparison has to be rejected as erroneous, as I have been arguing. For there is no such thing as animals, which are first of all only animals and then, moreover, something else. Therefore, the question whether the mindedness of the living being called "the human being" puts us into the animal kingdom or raises us above it is a category mistake.

*A **category mistake** is an inadmissible transfer of a field of sense to objects which do not occur in it. In our case, this means that mind, consciousness, soul (or whatever expression we want to use to point to this dimension of our life) are not directly detectable in the universe with the usual methods of natural sciences. Nevertheless, it would be another category mistake to infer from this that we ought to banish mind to a beyond, so that one could no longer explain how it leaves causal traces in the universe.*

This somewhat abstract consideration can be explained as follows. If one wants to explain why certain physically or biochemically measurable natural phenomena (such as a digestive process) occur at a certain place, it often involves a so-called action explanation. That is, one answers the question why someone has done something (e.g. eaten a vegetable dumpling) which is a crucial part of the explanation of why digestion is taking place. What one cites to explain the action (e.g. the person's vegetarian beliefs) is part of the total cause of the course of natural phenomena. Some natural phenomena involving us as minded beings (for example complex action formations such as a federal election that also leaves massive traces in the causal fabric of the universe) simply cannot be meaningfully explained if one disregards minded acts of self-determination.[119] Other processes, such as the digestion of a dumpling, occur largely independently of what intention someone may have had who consumed the dumpling. Therefore, it follows that the mind is not simply a part of nature. This does not mean that we are disembodied subjects, immortal souls. But it also does not follow from our mindedness that we are necessarily embodied or even identical with our organism.

Our self-image includes not only our mental abilities in the narrower sense but also those dimensions of our life which we attribute to being an animal. Thus, thanks to the findings of the modern life sciences, anonymously running biological processes and, therewith, our body as parts also belong to our mind. From the point of view of the mind, the body is therefore in the mind – and not, as many people think today (and even consider to be scientifically proven) the other way round. In general, the mind manifests itself in the exercise of a person's faculty of self-picturing. Human pictures are essential for what we are.

At first sight this certainly strange thesis can be understood better with a short excursion into **mereology**, the doctrine of wholes and their parts.[120] Logically considered, relations between wholes and parts are not limited to arrangements of physical things. The thought "2 + 2 = 4" also has parts (such as the numbers occurring in it), without the thought (or even the numbers occurring in it) thereby being physical. Napoleon is a part of European history. He is not just thereby presently a physical thing.

From a mereological point of view, there is nothing to be said against understanding something corporeal as part of something non-corporeal. My thought that it is drizzling right now in Zurich (where I type these lines) is something non-corporeal. In this thought, Zurich occurs as well as the drizzle. Granted, philosophy of language has been arguing for quite some time about the question whether corporeal things such as Mont Blanc are parts of thoughts or not rather linguistic or neural representatives of these things. However, it leads to absurdities to assume we cannot think directly about Zurich, drizzle, or Mont Blanc, but must instead make do with linguistic, neural or other such images of these things. If I could only think about the drizzle when I have a neural or other image (representation) of drizzle, then how would I even know drizzle exists? For I could not compare the drizzle itself with my representation of it without thereby again comparing only one representation with another representation. The logician, mathematician and philosopher Gottlob Frege pithily sums this up in his famous essay "The Thought: A Logical Investigation":

> If I do not know that a picture is meant to represent Cologne Cathedral then I do not know with what to compare the picture to decide on its truth. A correspondence, moreover, can only be perfect if the corresponding things coincide and are, therefore, not distinct things at all. It is said to be possible to establish the authenticity of a banknote by comparing it stereoscopically with an authentic one. But it would be ridiculous to try to compare a gold piece with a twenty-mark note stereoscopically. It would only be possible to compare an idea with a thing if the thing were an idea too.[121]

From the mereological point of view, there is nothing to say against the fact that the physical reality can be part of thoughts which are not themselves physical. To be precise, the impressive success story of the modern mathematized natural sciences, physics above all, speaks for this. Admittedly, nobody really knows why it is even possible to capture nature at least partially in mathematical models and to arrive at new insights into nature through pure mathematical research, e.g. in the context of theoretical physics. In this connection, the Nobel Prize winner in physics Eugene Wigner even speaks of an "unreasonable effectiveness of mathematics."[122] However, whatever the solution to this riddle may be, given our logical-mathematical abilities we are certainly able to explore the universe to an impressively large extent. Likewise, given the language of mathematics we are able to look infinitely far beyond the horizon of our minds.[123]

So if there are no good reasons in general (mereologically considered) against the assumption that bodily reality can be part of non-bodily wholes, why should we not bring this idea to bear when it comes to the relation between mind and nature?

And exactly therein lies a key to the realization of our animality. For as minded living beings our body is in many ways part of our mind, our picture of ourselves. This is of course familiar from psychology, psychiatry, and neurological and cognitive research, insofar as it is known that each of us has a body schema, i.e. an image of what our own body looks like. We therefore always perceive our body in a certain way and imagine how we look to others, how our voice sounds, etc. This picture of ourselves is always a distortion. Our self-image is prone to error and falsification, which everyone knows from the experience of seeing themselves in the mirror or in photos or hearing their own voice.

But besides this body schema, which is more like a picture of our body than our body itself, one can go further and say our body is itself part of our mind. This is shown in the fact that we can take care of our body and consciously work on it or refrain from doing so and let ourselves go. The human body is not simply a part of the environment of the mind but, as Ludwig Wittgenstein once put it: "the best picture of the human soul."[124]

Our bodies are not a dark side we carry within us because our organisms have been shaped by millions of years of evolution and therefore still exhibit deep stimulus–response patterns and basic emotions such as fear, aggression, affection, satisfaction, etc. Our body is not an 'animal within us.' It is not as if evolutionarily older layers of our organism – such as the brain stem or the microbiome, which inhabits our intestinal flora – control us, like animal instincts which we get under control with effort and hardship in a civilizing and semi-rational way. Rather, as minded living beings, we appropriate our body and make it part of our mind – until at the latest through illness and death it takes over and can cloud the mind to the point of complete extinction. The appropriation of our body by our mental picture of it also affects dimensions of our body which we do not exactly know. And many processes in our body actually run autopoietically, i.e. through self-organization of subsystems to which we have no conscious access whatsoever (think of the brain stem or the bacteria in the intestine).

That the body is part of the mind in no way implies we therefore have an immortal soul. Rather, the belief that we have an immortal soul is a possible answer to the question of what mind is. From the fact we are minded beings, however, it does not follow that we can survive our lives, so to speak, and continue to exist after the death of the body. We simply don't know whether physical death is the end of life.

PART II

SOCIAL FREEDOM AND THE MEANING OF LIFE

The proper function of man is to live, not to exist.
I shall not waste my days in trying to prolong them. I shall use my time.

Jack London

Biology alone cannot give us any information about the meaning of life. But this is not a shortcoming. For biology is not equipped with the concepts which would allow us to recognize in human life and survival any special rank, meaning, value or dignity.

From a biological point of view, we embody only one form assumed by life, and this is one form among many others – although no one knows today whether outside the Earth an evolution of living beings has also occurred resembling that on planet Earth, or whether life itself (if this is a uniform phenomenon at all) is a highly local and exceptional phenomenon of nature.

As already mentioned, there is still no binding definition of life but only more or less well-established criteria for its recognition, which are mostly based on the assumption that cells are the basic building blocks of life. However, this is already a limited view, because many cell types consist of subsystems (such as organelles) which cannot be denied life.

The idea that we live in the geological age of the human being, the Anthropocene, is misguided. For we are not even remotely as central to natural events as our current, undebatable self-endangerment through human-fueled global warming would have us believe. As living beings, humans are not biologically exceptional. In this part of the book we will see this does not mean we are only poor stardust or a "mouldy film"[1] on the planet, as the pessimistic, usually bad-tempered philosopher Arthur Schopenhauer thought. It is an error, more precisely a category mistake, to predicate the meaning of human life on biological parameters.

Part I dealt with the question to what extent the human being is an animal. It turned out the concept of animal contains a hidden anthropocentric projection. It says more about how we humans distinguish ourselves from other 'animals' (or even 'animals' as a whole) than about what animals are like in reality.

I have suggested that, in our handed-down self-determination as animals, we should orient ourselves even more to the life sciences than usual and recognize that, although there are innumerable living beings and life forms, they cannot be meaningfully subsumed under the ill-formed category of 'animal.' The range of life, which probably comprises more than a trillion species (including innumerable microorganisms) is far too complex to correspond to simple classifications such as plant, animal and human.[2]

In the following we will deal with the problem of the meaning of life. The question of the meaning of life is closely connected with the fact that we are mortal and vulnerable. Whether and how we understand human beings as animals is inseparably woven together with our evaluation of life. But is there any sense at all in asking about the meaning of life, or must we with metaphysical sobriety assume life is a more or less random succession of forms arising on our planet like a "mouldy film"?

Part II will also be about the connection between the question of the meaning of life and our ideas of bare survival. And it will turn out these questions are profoundly political, because they affect our ideas about the larger community. That biopolitics occupies a central place in the politics of the twenty-first century should have struck everyone since the beginning of the Covid pandemic.[3]

The meaning of life is not exhausted in mere organic survival. For survival can lose its meaning; and it is not always and for everyone possible to find it again. Life can become a tragedy, which it is for an inconceivable number of people to whom terrible things happen. Besides natural deaths and random accidents taking our lives, socio-economically produced suffering also affects countless people. Such partly natural, partly accidental, partly socially produced tragic suffering has no deeper meaning. It is painfully meaningless. Nothing justifies it.

Conversely, we humans experience it as fulfilling and meaningful when we contribute through our activities to the further development of our lives together with the lives of others. Working to make tragedies less likely, to ensure that as many people as possible can live and die in dignity, that we take care together through principles of social justice to ensure we live in societies that allow as many people as possible to find meaning in their lives without harming others – all this rightly seems meaningful to us. Against this background, I will unfold in the following the thought that the meaning of life is the **good life** and that this consists in our working together on human, moral progress.

The Basic Idea of Liberal Pluralism

The goal and meaning of our earthly life is that we delineate margins of social freedom which allow every human being to find meaning in life, whatever this may then consist of in the individual. This is the **basic idea of liberal pluralism**. In this context, freedom is not an individualistic concept, not a concept distinguishing us humans from one another or even positioning our interests against one another, but rather something uniting us. Liberal pluralism must therefore not be confused with unleashed neoliberalism, which sees individual freedom in the fact we have preferences we both pursue in loose strategic cooperation with others and enforce against the preferences of others. Rather, we humans are so intimately intertwined that no exercise of our freedom is ever completely asocial.

Liberal pluralism accordingly has an ethical foundation. The point of liberal pluralism is not to justify the currently prevailing, globally perspicuous and morally reprehensible division of labor in modern societies, in which wealth and resources are distributed in such a way that too many people suffer directly and indirectly from the consumption needs of rich industrialized nations. The good life requires that we work to ensure as many people as possible enjoy it. We feel the suffering of others because we humans are intertwined. This is one of the reasons why it is traumatic to experience a pandemic or even a war of aggression, even if you are not directly affected yourself.

The individual freedom to find meaning in life is, I will argue, essentially social. Most of what we like to do, after all, we do with others. The freedom to play tennis, kiss someone, drink wine, or read vegan cookbooks is not available to anyone without other people playing a morally relevant role. One should not cheat at tennis, one should only kiss people who accept this, one should not drink wine while driving a car, and one should not buy vegan cookbooks whose royalties benefit the media channels of Attila Hildmann (a famous German author of cookbooks and political lunatic) spreading agitation, xenophobia and anti-Semitism.

Human freedom is always social freedom. Even the freedom to withdraw and come into contact with as few people as possible is socially regulated, because it is not appropriate, for example, to disturb hermits in their mountain caves or to pressurize people who like to live alone into a forced marriage.

As seen in the first part, the human being is not only an animal. Rather we are specifically minded beings who are capable among other things of higher morality and explicit ethical knowledge. We will not convince bulls that incest is harmful to their family structure; we will not turn lions into vegetarians to protect the poor gazelles; we will not convince chimpanzees of feminism; and so on. By contrast, we expect each other as humans to work together for moral progress. True, different groups of people have different ideas about the content of moral progress. Across populations there are moral value judgments which are incompatible with each other at one point or another. But for all that, only humans are able to formulate explicit ethical value judgments and to raise them to error-prone and revisable knowledge claims. Moral disagreement only shows it is possible to get moral reality right or wrong.

This is not an objection against, but an argument for animal and environmental ethics. Because of our ethical knowledge, we are responsible for other living beings and our shared environment in a way that is not true for pigs, snakes, forests, fungi and bacteria. There is a categorical difference between the prosocial behavior of other living beings (which also cooperate with fellow species members and form symbioses across supposedly closed species boundaries) and human ethics. For the latter goes through historically complex processes of progress and regress, which are made possible by linguistic, institutional and cultural factors that cannot be meaningfully represented in the languages of the life sciences. The ethical structure of our life form, our ethical life, differs from the conditions of our survival, i.e. of the production and reproduction of our bodies. While ethics can contribute to a flourishing life, it cannot be reduced to a set of evolutionary functions.

The human being exists, so to speak, at the interface of mind and nature. We know facts about both mind and nature, and at the same time we know we do not know them in their entirety – not to mention that we cannot grasp the exact connection between mind and nature either. We do not know what consciousness is and which aspects of our mental inner life are ultimately identical with brain waves or neuron storms.[4] It is part of the meaning of life that we can never fully grasp it and make it the basis of our societies as a catalog of instructions and rules of a good life. In this respect, liberal pluralism is secular – that is, as politically neutral as possible with respect to specific religious and ideological views of the meaning of life – but this does not make it atheistic, postmetaphysical or anti-religious. Our conception of life is never completely non-metaphysical and thus simply secular, because our self-portrait as an animal is based on valuations that go far beyond what we know about ourselves from life science. Therefore, biologism and creationism, which are hostile to each other, are both metaphysical positions that exist as extremes within liberal pluralism.

Biologism is the thesis that the meaning of human life in all its social, historical, cultural and institutional manifestations and therewith our social freedom can ultimately be traced back to the fact we are animals whose life form can be completely deciphered by biology. This view is just as ideological (and thus wrong) as its extreme opposite, **creationism**, which believes that all life processes are directly caused and controlled by God, so biology is based on fundamentally false premises.

However, even liberal pluralism is not nearly as modest and metaphysically secular as it presents itself to be. We cannot simply assume that everyone determines the meaning of his or her own life. Self-determination is a demand on the structure of freedom that has historically emerged from the Enlightenment. The Enlightenment argues we can exercise our self-determination, our autonomy, independently of which metaphysical worldview is ultimately correct, but thereby overlooks the fact that this very conception of self-determination is also based on a claim to truth. Liberal pluralism raises an ethical claim that determines all of our social life, for it calls upon us to recognize the self-determination of others as legitimate, even if it goes against the grain. Accordingly, it is not modest or neutral, even if, in my view, it rests on a correct ethical foundation.

Nor is it entirely secular, because it is based on metaphysical views about how human self-determination (i.e. autonomy) works. This has recently been argued by the famous political scientist Francis Fukuyama in a seminal book on liberalism.[5] How our autonomy fits into reality cannot be answered exclusively in terms of natural science and technology, as he argues; there are too many unanswered questions for that (such as what consciousness is and how free will is compatible with the laws of nature). Nevertheless, we are justified in holding on to our autonomy and defending liberal pluralism as a set of values precisely because it is ethically grounded in the concept of human dignity and thus the infinite value of each person in their individuality.

Liberal views, which are the basis of liberal democracy, are – as, for example, the Canadian philosopher Charles Taylor has pointed out in a series of seminal books[6] – the heritage of a multitude of religious traditions. Liberal pluralism therefore has room for the development of metaphysics, that is, for concrete ideas about which areas of reality cannot be explored empirically but can only be grasped through specu-lative concepts.

Enlightenment processes are morally ambivalent because, on the one hand, they made moral progress possible but, on the other hand, they were connected with colonial structures of exploitation, misogyny (women's suffrage is a very late invention of modernity) and implicit and explicit racism, which critics of the Eurocentric bias of modern Enlightenment have pointed out. But these phenomena are not inherent in the ideas of the Enlightenment, especially since they occurred and continue to occur in much sharper form before the modern European Enlightenment as well as utterly independent of it. But the Enlightenment needs further Enlightenment – a process that is not completed and indeed cannot be completed, because humanity is always confronted with other challenges to which we can respond with moral correctness or incorrectness, i.e. somewhere on the spectrum between good and evil.[7] Commitment to the Enlightenment conception of moral progress differs from the idea of an end of history. Liberal pluralism is not some final stage of human development but an attempt to leave history open, to correct ourselves in light of novel challenges arising through social change. Moral progress is not mere social change but positive social change whose direction is set by ethical insight.

Liberal pluralism is based on a conception of the meaning of life which allows social freedom to be given a political shape. The goal of social freedom is to allow and even promote as many life designs as possible. Each should be able to go through the search for meaning in his or her own way, without thereby harming others in their abilities to develop meaning in life. Liberal pluralism creates scope for the development of social freedom with the ethical goal of promoting moral progress.

The balance this requires between an open variety of life plans is expressed in the modern triad of freedom/equality/solidarity, which goes back to the French Revolution. This triad describes an ideal of values that we have to realize again and again in concrete situations of action and decision-making. What concrete situations (historical contexts) require for the realization of the triadic structure of social freedom cannot be gleaned from any formal conception of freedom/equality/solidarity but has to be figured out on the spot, as it were, by considering how non-moral and moral facts hang together in light of what we, humans, know in a given situation.

On the one hand, this presupposes we make value judgments, i.e. we decide who we are and who we want to be. On the other hand, these value judgments are revisable and prone to error, because we never know enough about ourselves, others and nature to make error-free judgments in complex situations.

Any number of individual judgments are admittedly true and in this sense error-free. Right now I judge that I am writing these lines in an airplane from San Francisco to Frankfurt, which is true. I also judge that I see my screen in front of me; that I will finish writing this sentence within the next minute; that I have ten fingers; etc. We never miss reality altogether and are therefore quite capable of recognizing it as it is. In complex situations, however, too many interrelated judgments are necessary to guarantee we recognize the situation as a whole and thus in its dynamic development. Therefore, our cognitive life is always *noisy*, messy and error-prone in an unpredictable way, as Daniel Kahneman, Olivier Sibony and Cass R. Sunstein have shown in their book *Noise*.[8]

But one must not use this complexity as an excuse to deny the existence of objectively valid value judgments. On the contrary, complex ethical problems are indications there are objectively valid values that put us in difficult decision-making situations. Complex ethical problems, which we all face at some point, presuppose there are shared and largely obviously correct value judgments that are bases for dealing with complex problems.

Life itself is complex. Therefore, it should come as no surprise that the meaning of life is not a simple matter.

The History of Life

The concept of life has a long and interwoven history. It even has several histories. In Africa and India, in China or in Western cultures, life was and is thought about in very different ways. For example, many cultures consider almost everything to be alive, which can include stones and celestial bodies – as described, for example, by the aforementioned Aboriginal thinker Tyson Yunkaporta in his book *Sand Talk*. The idea that only those entities are alive which consist of cells is fairly recent in the history of humanity, since we have known that cells exist only for a few centuries. The view that life is exclusively a property of cells or of systems consisting of cells has not really been substantiated until the present day; it is a modern dogma, which was pointed out to me by the late Humberto Maturana during a visit to Santiago, Chile, in spring 2018. The biologist and philosopher, who had become famous for linking the living character of the cell with the concept of **autopoiesis**, i.e. the self-generation and self-maintenance of a system, concentrated in the last years of his life on researching the biology of love.[9] I remember very well how he entered the room where I was waiting for him at his home and immediately told me he had come to the realization that, ultimately, the entire cosmos is alive and it was a mistake to regard the autopoietic structure of the cell as the decisive, necessary criterion of life. Rather, the connections between systems – and thus also love as an outstanding, valuable connection between several living beings – are decisive for how we should understand life.

However, not only the concept of life but even life itself has a long history. We do not know its exact beginning.[10] There are only more or less well-founded speculations about how in our universe material-energetic systems, which were not yet alive, developed at some time into systems which could be classified as alive. Perhaps the question is also misconstrued; that depends on which systems one considers to be alive. If indigenous groups of people who consider stones or galaxies to be alive are correct, the idea that life arose at some time on our planet would be wrong.

In any case, the history of the building blocks of life goes back at least to the formation of carbon, without which there would be no life form we know on planet Earth. This process alone raises physical puzzles, which among other things refer to a mathematical-physical problem in the calculation of the so-called Hoyle state. This state is named after Sir Fred Hoyle, who incidentally also coined the expression "Big Bang." The Hoyle state is about

> a high-energy form of the carbon nucleus. It is the mountain pass over which one passes from one valley to another: from three nuclei of the gas helium to the much larger carbon nucleus. ... Without this type of carbon nucleus, life would probably not have been possible.[11]

The more detailed investigation of the structure of this state has even moved metaphysically abstinent physicists such as my Bonn colleague Ulf-G. Meißner or the Nobel Prize winner Steven Weinberg to seriously consider a weak anthropic principle in physics research.[12] In general, the anthropic principle (from Greek *anthrôpos* = human being) states the human being is essential for the knowledge or composition of the universe. That means we are not a cosmic accident.

---

*The **weak anthropic principle** states measurements which physicists detect in the universe depend on the fact that they are performed by living beings which are carbon-based. Our factual presence in the universe is reflected in its basic structures because otherwise we would not exist. Therefore, the universe cannot be a place which is completely alien to us. It is, rather, recognizable and explorable for us at least insofar as it is based on the principles from which cognizing living beings, who are capable of doing physics, have at some point originated.*

The physicists Barrow and Tipler argue this weak principle cannot be disputed, because it merely says the universe which we can observe is the way we find it to be because we can cognize it. Otherwise we wouldn't exist, after all. And that includes the formation of carbon under complex conditions. Cognizing living beings which are not carbon-based would probably see a different situation from their own point of view.[13]

Of course the question cannot be answered as to whether one can conclude something about the significance of cognizant life for the universe from the complex physical preconditions of the origin of that form of matter without which there could be no carbon-based life. Moreover, the weak anthropic principle alone does not tell us whether a **strong anthropic principle** can be assumed, according to which the universe necessarily runs towards its self-knowledge in cognizant living beings like us. We can't decide that, because we don't know any other universes and therefore we don't know whether in all universes cognizant life can or even has to arise. If there are universes in which no cognizant life develops, its existence is not necessary.

Under which conditions life develops is a different matter, because ultimately we do not have sufficient scientific knowledge at present to trace the history of the origin of life. I touch on these delicate issues here only as examples for the range of our factual far-reaching not-knowing, which concerns the physical and biological bases of our embodied existence.

The fact that both the concept of life and life itself have a history makes the situation even more complex than the question concerning the origin of life. For, depending on how we determine the concept of life, we will look for other beginnings of it and will make a find in other ways.

As if all this would not be reason enough to be quite modest about life epistemologically, there is the persistent additional problem with the phenomenon of life: we simply do not know today whether there is life on other planets or even outside of planets. So we also do not know whether the general laws of nature now known, laws which apply everywhere in the universe, imply that under similar conditions life always arises (e.g. on all habitable planets).

In short, both the concept of life and life itself pose innumerable problems that have not been solved today. But it does not follow that life itself is a riddle or a mystery – only that there are innumerable things we do not know about life and its history. We also do not know altogether what we do not know about life, as there are only some gaps in our knowledge we are aware of. Therefore, regarding life, epistemological modesty and ethical humility are required. For life is clearly an important source of value.

The Idea of Life

Although life is in different respects a complex matter about which our knowledge is limited, at least one history of life is revealing in our quest for the meaning of life. This history is closely interwoven with the idea of developing a science of nature in the first place – an idea which, as far as we know, goes back to the ancient Greeks. They invented not only physics and mathematics but also biology as a science, because they wanted to know how the various natural phenomena they observed were connected. They were eager to understand reality as a system (yet another Greek term), which means a hanging together of observable phenomena.

In doing so, they came up with the unquestionably ingenious idea that natural phenomena follow certain patterns arranged in time in such a way that one can determine regularities in the natural processes. These in turn can be grasped logically and mathematically. This idea, familiar to all students of the sciences today, was a rather radical innovation at the time.[14]

For our context, it is significant that the Greeks considered as life those phenomena which consist in the fact that a general pattern, a genus (a *genos*, Latin: *genus*), repeats itself in individual cases. Thus, for example, the approximate appearance of a horse in contrast to an elephant. The Greek word for genus is connected with the word for becoming (*genesis*). A genus is something that persists in the processes of nature, a pattern that remains the same in change.

Now it is true that in reality there are not genera. There are only individuals that have in common whatever the concept of genus captures. According to today's view, this can be determined by molecular genetics.

(Genetics does not contain the term genus by chance.) If we look for generic patterns on the level of cellular blueprints, we can compile the family of primates, for example. This leads to our current distinction between four living genera: gorillas, humans, orangutans and chimpanzees.

These genera belong to the same family on account of their lineages, which are determined genetically. In concrete terms, this means certain genetic patterns are recognizable in each individual belonging to a genus, thanks to which we can assign the individual to a genus.

Species (Greek *eidê*, Latin *species*) form a suborder within a genus according to the Greek original idea. The specimens of a species have a similar look, which is a possible translation of *eidos* or *species*. All humans resemble outwardly all humans, all chimpanzees all chimpanzees. At this level, the concept of species corresponds roughly to what is now called the phenotype, i.e. the outward appearance of a living being. Species further divide classes of living things that resemble each other (genera). They are more fine-grained.

Of course, the zoological classifications are difficult in individual cases and discussed in the life-scientific disciplines, which deal with the question of how to arrive at an adequate classification of nature in view of the methodical state of research. In our context, however, it is not a matter of formulating a biological thesis in the strict sense of the word; others are called upon to do that. It is much more important to understand that the idea of life is not limited to what we can recognize with life-scientific methods, e.g. those of molecular genetics.

Life is by no means only a life-scientific topic, or even only a biological one. And this was exactly the ancient Greek insight that leads us to the *idea of life*, as it was called later by Georg Wilhelm Friedrich Hegel, who has again and for some time now taken center stage in the current discussion on the philosophy of life.[15]

Hegel argued that life is a property not only of our bodies but also of our minds. In the chapter dedicated to "Life" in his *Science of Logic* (a work I can only recommend to those who want to spend the next ten years with it), Hegel reinterprets the Greek discovery that life is not limited to biological life. Crucial here is his realization that we conceptualize not only organic systems as "living" but also quite different phenomena. And in fact it is true that the Greeks thought of thinking processes as exemplary when they discussed the concept of life.

Life in Greek is *bios* (hence biology) or *zoê* (hence zoology). Living beings are called *zôa*, which is derived from the second word. I do not mention this here only as a scholarly comment. For it is important that Aristotle understands life as an "activity of understanding" (*energeia nou*),[16] i.e. not primarily or exclusively as a process, thanks to which elementary forms of life (e.g. cells), as formed organic carbon-based matter, assert themselves against the environmental pressure of entropy, i.e. – to put it simply – the dissolution of their structural patterns.

You have to let this thought roll off your tongue, so here it is again in somewhat smaller steps: *organic chemistry* studies chemical compounds based on carbon. These are found in all living things known to us. All living beings known to us distinguish on a fundamental level (paradigmatically on the cellular level) between themselves and their *environment*, as the biologist and philosopher Jakob Johann von Uexküll called it (see p. 178). Uexküll coined the term environment in 1909 in his book *Umwelt und Innenwelt der Tiere* and thus became a pioneer of ecology. At a conference on the philosophical foundations of neuroscience, which took place in November 2021 in Cáceres, Spain, the British zoologist and neuroscientist Matthew Cobb explained that each individual cell, and thus also each individual nerve cell, has an environment to which, above all, other nerve cells connected with it belong.[17] Accordingly, it is anyhow not correct that there is *one* environment. The term environment is relational; it designates the events and structures relevant for a living being and does not designate an enormous place (nature, the universe, the cosmos) where everything takes place.

Living, carbon-based systems are open, i.e. they exchange substances with their environment. In this way, they maintain their structure and resist entropy, which provides a partial answer to the question "What is life?" This was pointed out in particular by the famous quantum physicist Erwin Schrödinger, who introduced the concept of negentropy (i.e. the active generation and maintenance of structure, e.g. through diet or exercise, a term needed for the physics of living things) and thus had a major influence on molecular biology.[18] Entropy, which is derived from thermodynamics, ensures living things would dissolve much faster than they would like absent negentropy. We know entropy from everyday phenomena, such as the fact that we can easily pour milk into coffee but it is much more difficult to reverse this process, i.e. to turn a milk coffee back into a black coffee. In addition, every coffee gets cold faster than you would like, because it is connected to its environment and has to give off heat without being able to generate heat itself. Our coffee is not alive, as the cells of the coffee beans no longer reproduce themselves, and is therefore mercilessly exposed to entropy (which is an advantage, because otherwise it might be quite sadistic to drink living coffee).

The difference between self and environment thereby produces an elementary form of normativity (i.e. the difference between right and wrong behavior), as the eminent American philosopher Robert Boyce Brandom has shown in his 2019 masterpiece on Hegel, *A Spirit of Trust*.[19] Life forms distinguish between what is good for them (food) and harmful substances they should not acquire. Thus, already on an elementary level of life, an evaluative difference arises between a self and its environment, from which the self obtains nutritive substances for its metabolism. In his philosophy of nature, Hans Jonas already inferred from this that our selfhood, our subjectivity, is an expression of life processes.[20]

At this point, one might be tempted to take these basic properties of the life forms we know as very distant biological ancestors of our more complex social, ethical and political value judgments and, thus, ultimately to biologize them. But this would be a category mistake, i.e. an inadmissible transfer of forms of thought and facts, which are valid in one field of sense (that of biology or zoology), to another field of sense (ethics, politics, or history), in which they are not valid in the same way.

Objects always appear only in contexts, in fields of sense. They appear in a certain way, i.e. in qualitative and quantitative arrangements.[21] Objects are basically countable; they form units. There are individual books on my desk, there is a large screen in front of me, there is a keychain to my left. I could count these objects. That is a quantitative arrangement. But these objects also mean something to me. They have qualities for me. They are arranged in a more or less meaningful way for me. We never experience reality only quantitatively, i.e. as an accumulation of objects, but always qualitatively as well, in color, sound, taste and other sense modalities. Some fields of sense can be described and explored physically (e.g. the field of sense of the elementary particles or that of the galaxies); for other fields this would be nonsense. It is impossible to reduce all fields of sense to quantity; there are qualities which can be described mathematically (e.g. as vectors), which does not mean this representation reduces them to measurable quantities.[22]

This brings us back to Aristotle and Hegel. For they realized it is a category error to reduce life to organic processes, i.e. in this case to identify life with organically based processes which resist entropy as open systems. For in this way one does indeed obtain some basic concepts of the life of mind – such as self and non-self or an elementary normativity, i.e. a difference between something that should be (food) and something that should not be (pollutants). But it is not successful at understanding living processes that cannot be meaningfully described in the language of organic chemistry.

To living processes which are not chemically describable belong the difference between a self and a non-self as well as the further difference between whatever one is not oneself (a general non-self) and that which is another self, an *alter ego* or a you.

In this context, an ego is more than a self. We experience ourselves not only as a self embedded in an environment, from which it also differs, but also as a more or less rational control center of our life that pursues intentions and has a biography. This dimension of human life goes far beyond anything we attribute to cell cultures such as so-called biofilms, in which microorganisms of various kinds cooperate socially.[23] However important these may be to the processes of living things on our planet, they do not pursue rationally articulable intentions, even if they may be infinitely more persistent evolutionarily than more complex organic life forms.

In this context, Brandom has shown in his aforementioned book on Hegel that the existence of an elementary biological self already produces norms. A self (say, that of a unicellular organism) biochemically differentiates between itself and others and can thereby move in an environment and search for nutrients. An ego, on the other hand, can turn this circumstance, this orientation to norms, into a norm. Therefore, we have an understanding of our normative orientations as well as an understanding of the normative orientations (desires, hopes, values, goals, etc.) of other persons who are also capable of living a life in light of an idea of themselves. We make explicit our value orientations in language, institutions, art, religion, culture, etc., as Brandom points out elsewhere.[24] The social practices involved far exceed the production of goods for our survival. Society is therefore not just a "system of needs," as Hegel called it, i.e. something that can be organized economically and planned more or less well.[25]

To Live and to Survive – The Basic Form of Human Society

When we experience our life as meaningful, we have an experience involving other people. A meaningful, happy experience or a significant episode of our life (the new dream job, the great love, the birth of a child, the successful performance of a long-practiced piece of music, the experience of divine presence, etc.) goes beyond our individual experience. Even the lonely walk in the forest or a meditation, in which we try to reflect on ourselves and our senses without judgment, remain related to other people. Someone has taught us how to meditate, or we meditate (for example, in a Buddhist monastery) with others.[26] And the fact that one finds a solitary walk in the forest restorative and meaningful is also something one has learned from others at some point, for example, by having experiences in a forest club in elementary school, in kindergarten or with the family, which one recalls later in life, directly or indirectly, in the form of a walk in the forest. Those who consider the meaning of life to be something God has given, and thus as something to which they dedicate their lives, base this on a social experience of dealing with a divine You, celebrated by their religious community.

The human being is a social living being.[27] Our sociality is not something we acquire at some point, usually quite early in life. None of us would be able to read these lines if we had not already been in a social relationship with other living beings (first of all our mother's womb, in which process many non-human life forms are involved). The basic neural and other building plans of our organism are formed in utero, which is why we leave the womb with a rather complex equipment and personality – which all parents and educators can confirm from their own experience. At the same time, our individuality as an organism is itself quite complex. For instance, we all have individual immune systems, and these distinguish us from other people even more precisely than our fingerprints.[28]

Our organic structure, and thus our survival, is consequently already a social product, i.e. the result of socially coordinated processes in the womb. The mother must feed or be fed for the fetus to form a nervous system; this requires division of labor. And we must all be socially conceived in one way or another (usually through intercourse or some other form of socially orchestrated fertilization of an egg).

So far I have spoken of life forms in order to cover all living beings as far as possible and to keep away the confusing concept of 'animal,' which, as described in Part I, is a human, all-too-human projection. At this point, however, I would like to introduce another difference, which is important to grasp the meaning of life (in a sophisticated sense).

---

*Organic life, which the life sciences investigate successfully and often for the benefit of humanity, is defined by the **form of survival**. Medicine, for example, deals with the form of survival of organic systems (a heart, the skin, or a whole human being or other living being). It tries to preserve, support or even kill organic processes (e.g. a tumor, which is an unwanted body part, a form of survival we want to eliminate).*

But medicine as a rule (with the exception of psychosomatics, psychiatry, etc.) does not explore the life of the human mind (not to mention the divine mind, if it exists, which will be neither denied nor asserted here). The life of the mind includes the aesthetic experience of a symphony, the reading of Dante's *Divine Comedy*, a wild evening at a club, federal elections, or the activist engagement against political slowness in the fight against human-made climate change, to name only a few things among those that are meaningful to people.

Let us call all this, in contrast to survival and thus to our form of survival, our *form of life*.[29] Our form of life is the life of mind which is certainly related to our form of survival but cannot meaningfully be reduced to it. Since Aristotle, we also speak of our "second nature," which is however misleading, as the mind is ultimately not only an extended area of nature but categorially distinct from it.

At this point it is important to dispel a suspicion before it takes hold. The fact that mind and nature are *categorially* different from each other means the mind is not an ethereal ghost. The mind does not belong to the universe in the same way as photons, nuclear forces, gravity or neurons. It is not in the target system of physics. The fact that the mind is not an object of physics (but of metaphysics) does not mean it is a hidden part of nature (a secret ingredient in the development of the universe, some ethereal dark energy, as it were).

This is precisely what is overlooked by **Cartesian dualism** (going back to René Descartes), which is rightly criticized in this connection and assumes there are two substances in nature, i.e. two kinds of things – the extended, material bodies and the immaterial souls or thinking substances, which somehow (it is just not known how) causally interact.

This is not to say the mind is somehow mysterious or cannot be studied scientifically. Depression manifests itself in serotonin levels, joy in endorphin levels, and so on. This is what molecular psychiatry studies in detail. But the material manifestation of subjective states is not identical with these states. One measures serotonin and knows this correlates with subjective states. But what one measures in this way from the observer's perspective is not identical with the subjective state. Otherwise one would literally be able to see a subjective state, and that would be nonsense. One never sees one's own seeing. One sees, say, an apple tree. The subjective state of seeing is not itself visible. Of course that doesn't mean it's unreal.

Mind [*Geist*] is the subject of research in the humanities [*Geisteswissenschaften*] and social sciences, as well as of our everyday social, political, aesthetic or religious experience. To acknowledge that mind exists and that it can even be captured in mathematical models (such as those developed by economists or quantitative sociologists) is therefore not mysterious in any relevant sense.

The human mind is the source of the meaning of life. Meaning does not boil down to mere survival.

We can distinguish between our form of survival and our form of life, which is always also an affair of the mind. Both our form of survival and our form of life are constitutively social. In order for a human being to survive long enough to be able to exist alone, if necessary, like a person stranded on an island or a character from one of the countless recent post-apocalyptic films (as the last human being), we must have been raised by other human beings. Johann Gottlieb Fichte sums this up: "The human being becomes a human being only among human beings."[30]

Do We Want to Live Forever?

Mere survival cannot be the meaning of life. Whether people still find their lives meaningful if, at a very old age, they outlive their family and friends depends on many factors and is assessed differently depending on the person. Philosophically, it becomes even more difficult if we imagine we would be forced to live a (more or less) earthly life for all time, like the Flying Dutchman or some cursed ghost who has to live forever.

In recent philosophy, the issue of the meaning of eternal life has been hotly discussed since an essay by Bernard Williams. Willams draws on a fictional character by the Czech writer Karel Čapek (who, incidentally, also coined the term robot). Čapek's comedy *The Makropulos Affair* (first performed in Prague in 1922) centers on a character by the name of Elina Makropulos, who drank an elixir of life over 300 years ago and thereby became immortal by earthly standards.[31] The question now raised by this familiar human fantasy of earthly immortality is whether an eternal earthly life would be meaningful at all or whether it would not rather be a tragedy for those involved. Let us call this the **Makropulos thought experiment.**

The negative view according to which an immortal life would be a tragedy and that we should therefore be content with our own mortality, because without it, without death, there would be no meaning to life, has recently been advocated by the Swedish literary scholar Martin Hägglund, in his book *This Life: Secular Faith and Spiritual Freedom.*[32] It is true, of course, that we choose among options for a source of meaning in light of our knowledge of our own finitude. But it does not follow that death, or our attitude towards it, is the primary let alone the only source of meaning, such that all meaning would dissolve into nothing if we were immortal. Hägglund takes up an old topos of existential philosophy at the heart of Martin Heidegger's *Being and Time.* In his magnum opus, Heidegger describes humanity's mode of being as "being-toward-death."[33] If he were right, everything that actually matters to us (our friends, our favorite things, etc.) would have significance only because, in truth, we always already experience our lives in light of our death. Heidegger thus radicalizes the baroque motif of *memento mori* (remember death) and ultimately asserts we could not even furnish our garage with a workbench without thinking (rather unconsciously) of our death.

This train of thought results from the fact that we would certainly become rather bored by our earthly lives at some point if we had to stay here forever. Besides, it would be perhaps particularly terrible for the immortals among us to experience again and again how the mortal loved ones disappear.

But these results are, on closer examination, only articulations of a one-sided emphasis on one or the other aspect of the Makropulos thought experiment, i.e. the question whether an infinitely long earthly life would be desirable and thus meaningful. After all, how do we know that after a few centuries one might not get used to spending an average of fifty to eighty years with the same friends and family before then taking on a new identity and beginning a new life? And what if everyone were immortal? Why would it be boring to work with eight or nine billion immortals to build a paradise on Earth? It is more probable that some immortals would find this good and others bad – which is hardly surprising, as in this thought experiment we're dealing with people who are immortal but nonetheless remain human.

In this connection, Friedrich Nietzsche is probably right, as he presented his own, cosmic version of the Makropulos thought experiment in *Thus Spake Zarathustra*. Nietzsche imagines that the universe always repeats itself in a cyclical way, so we all go through an "eternal return of the same." Nietzsche now says that it is the real task of a "revaluation of all values," i.e. a new ethics, to see such a scenario precisely not as horror or tragedy but as a source of meaning and to live in such a way that everything one experiences could also be lived through an infinite number of times. To be sure, this optimistic or heroic vision does not convince those to whom terrible things have happened. For far too many people of the past, present and future it would certainly be no consolation if they had to go through their entire life in an endless loop (at any rate without noticing it) an infinite number of times.

***

Long story short: the question whether our life is meaningful only in contrast to death is a question that cannot really be answered, at least by recourse to literary fictions and other thought experiments. On closer inspection, it is nonsense and even ethically irresponsible to claim death is the source of the meaning of life. Our own death, much less that of a loved one, is far from being a source of existential meaning, let alone the primary one. Death is not a source of meaning but rather, as the radical end of a life, the direct opposite. Even if it can be a redemption from earthly torments, the meaning of life does not point towards death. It gives us no orientation at all.

Nietzsche imagined that religions only represent an immortal life in a paradise (or the final exit from the cycle of rebirths) because they despise our earthly human life and want to flee into cloud cuckoo land. Such a wild claim fares no better than the idea that death is the meaning of life. In order to support such a critique of religion as a flight from finitude, it is usually assumed an eternal life in paradise is boring. People do not seriously want to celebrate an eternal Christmas or Holi with their families but are secretly glad life will end at some point – a theme the amusing American television series *The Good Place* explores in fifty-three episodes. The immortals in paradise long for extinction in the end and decide in favor of it.

But, again, none of us knows whether there is such an eternal life after earthly death and what it might look like. Certainly, it would be more like hell for many if we were to find ourselves on planet Earth or in the TV series *The Good Place* after our clinical death, although this is probably also a question of character, since it might be more boring for the viewers than for the characters.

The widespread thought experiment which asks us to imagine an eternal continuation of our life in order to discuss the question whether immortality is desirable, and thus the source of the meaning of life, fails because we use our hopes and fears to answer the question and embellish the thought experiment in each case in such a way that the result is already certain. Whether one wants to live forever or not says more about one's own overall psychological state than about whether eternal life would be objectively good or bad. Eternal life is just as good or bad as finite life. The quality of life and its meaning do not depend on some time span or other.

Incidentally, *The Good Place* is ultimately just a US sequel to what the French writer and philosopher Jean-Paul Sartre wrote in his drama *No*

Exit, which was first performed in 1944. In this drama, three people find themselves in a hell after their death, which consists in the fact that, from now on, they must always live together, from which Sartre derives his famous dictum "Hell is other people."[34] Sartre, like many other existentialists in the wake of Heidegger and Nietzsche, was convinced that death is a source of the meaning of life, as he explores through this thought experiment. In doing so, he made this assumption, like Heidegger, because it radically throws us back on ourselves and confronts us with the fact that we die in metaphysical solitude. While Sartre also distances himself from Heidegger because he sees death only as *one* element of our facticity, i.e. our basic existential endowment, he maintains that we can only find meaning as individuals radically facing their own death. Simone de Beauvoir, who (unlike her companion Sartre) assumed our existence is fundamentally social, rightly argued against this individualism. We neither live nor die alone; all we do and think is directly or indirectly social.

In any case, Sartre's drama has no evidential power whatsoever to prove universally, in a way valid for all people, that eternal life would be meaningless. His drama is and remains a fiction. We cannot conclude from it whether death is the decisive source of meaning for life. Such a conclusion says more about whether we ourselves would like to be immortal. It says less about whether this is universally desirable.

The truth is we simply do not know what will become of us after our clinical death. The only certainty is that our cells will dissolve after a while and our earthly body will not outlive its death. But it doesn't follow that there will be no resurrection of the body, nor that we will not have an immortal soul, to be granted a highly desirable but in this world unimaginable eternal life. Likewise, natural science cannot disprove or exclude the belief in reincarnation, which plays a central role in many religions, because natural sciences always merely describe, explain and partially predict how certain processes and structures develop in the universe. It is never possible to deduce from this that there exist only such processes and structures as are investigated by the natural sciences. Whoever trusts the natural sciences to have abolished transcendence is a victim of an intemperate overestimation of the human power of cognition. Here it is perhaps salutary to recall the great physicist Erwin Schrödinger, who co-founded quantum mechanics. In his influential book *Mind and Matter* he explicitly argues for the existence of a unified, all-encompassing cosmic consciousness, thereby echoing Hindu thought traditions. This is not to claim that Schrödinger was right. But it does prove that the best possible knowledge of quantum physics in its connection with thermodynamic processes does not lead to the realization that there is only material-energetic reality and nothing beyond. In short: physics by its very nature differs from metaphysics and can therefore make neither positive nor negative statements about transcendence. The same is true for the life sciences, which will never be able to prove we are identical with our body. So it is merely modern hubris to spread the fairy tale that we finally know that death is the ultimate end. We simply do not know such things. To claim that death is the meaning of life, or the source of the meaning of life, is therefore a baseless assumption, even if widespread. It certainly cannot be substantiated by reference to fictional thought experiments. Our assessment of the question whether eternal life is meaningful rather expresses our attitude towards this earthly life of ours.

The Meaning in *Life*

The philosophical discussion about the meaning of life has taken an interesting turn since the 1980s. Susan Wolf has made a particularly important contribution.[35] Instead of talking about the meaning *of* life, she has shifted the perspective and asks about the meaning *in* life. In order to deduce what meaning actually consists of, she suggests that we first become aware of which experiences and attitudes we consider meaningful *in* our lives, before we deal with the – in Wolf's view too big – question of *the* meaning of life.

Here it is crucial to note first of all that different people experience very different things, situations and actions as meaningful. Some people like to fish, others are absorbed in their nursing profession; some like to travel to the mountains and work towards their vacation every year, while others devote their lives to computer games; some are family people and can't imagine anything better than raising children, yet others couldn't bear to live with children because they would rather write postmodern novels in solitude. The list could be extended at will, which shows there is no limited set of human activities or preferences from which the meaning in life could be derived for each individual.

It is an achievement of modern liberal pluralism to recognize this and accordingly to set up the political circumstances of a state in such a way that, so far as possible, we do not define any particular way of life (such as a religious way of life or the role model of the nuclear family) as the standard around which we organize our institutions.

Unfortunately, our institutions have not been sufficiently liberal. That it has taken so long to make same-sex marriage a reality is regrettable; that it exists in the meantime is a sign of moral progress. In a liberal democracy, who raises children or not (and with whom) should be assessed only in light of the question of whether the legal guardians can adequately care for the children. Institutions exist for this to provide supervision for people who are granted or denied custody rights.

It is therefore legitimate to examine our institutions and public debates for their liberal pluralism and, if necessary, to criticize and correct them in order to overcome one-sided ideological imbalances – which also includes the widespread criticism of liberalism, which says that it is inherently designed for neoliberal relations of exploitation and thus incompatible with social freedom. As the social philosopher Lisa Herzog and the historian Ute Frevert have shown independently of each other, a contemporary liberalism is conceivable that connects the market economy back to moral values.[36] In my view, this means that the institutional (above all legal and political) framework conditions of a society's production of surplus value should always be oriented to moral facts, so the logic of economic action can be used not only for profit but also for the advancement of the common good. This is, after all, the point of taxes, which are not an involuntary tribute to the state but a contribution to securing and expanding our social freedom. In a liberal democracy, the interaction of the state and the economy takes place within an ethically based framework of values. The neoliberal view that the economy should only serve itself and that the state's role is at best to defend free markets against external threats overlooks the fact that there are no markets without people, whose freedoms are always already social. In fact, many entrepreneurs behave in such a way that moral standards also play a decisive role in their economic activities, which is due to the fact that their companies are part of liberal democracies and are not required to provide funds for autocratic rulers and despots to oppress the population. So-called neoliberalism – which includes an influential economic school of thought and a world-historical practice associated in the West primarily with Ronald Reagan and Margaret Thatcher – is unfortunately anything but liberal in the final analysis. We are witnessing this once again not least in Russia, where, after the collapse of the Soviet Union, neoliberal reforms have created the type of oligarchy that transformed former state enterprises into kleptocratic systems. In this case, the supposedly free market is destroying the freedom of many people. But this is not simply because of the market economy and certainly not because of liberal democracy, but because of a radical decoupling of the state from the economy, which cannot be justified ethically by anything. Neoliberalism regards the state only as a system for securing the production of surplus value and not as an actor which should always contribute to moral progress through its democratic networking with society as a whole.

A crucial insight of liberalism is to recognize there is nothing that really makes every person happy in the same way. This becomes even clearer when – as is unfortunately common – one thinks of people as mostly only those more or less healthy adults who are entitled to vote. As almost always happens, it is easy to overlook the children and the sick, who probably have quite different priorities than the majority of the readers of these lines. Just remember what was important to you when you were seven or eleven years old.

Nevertheless, it does not follow from this consideration that there is nothing common to all experiences of meaning *in* life. Therefore, Susan Wolf proposes a concrete conception of *meaningfulness* that identifies a basic trait of meaning *in* life to which all people can relate. Let us call the following proposal **Wolf's love thesis:**

According to the conception of meaningfulness I wish to propose, meaning arises from loving objects worthy of love and engaging with them in a positive way. … "Love" is at least partly subjective, involving attitudes and feelings. In insisting that the requisite object must be "worthy of love" however, this conception of meaning invokes an objective standard. … according to my conception, meaning arises when subjective attraction meets objective attractiveness. Essentially, the idea is that a person's life can be meaningful only if she cares fairly deeply about some thing or things, only if she is gripped, excited, interested, engaged, or as I earlier put it, if she loves something – as opposed to being bored by or alienated from most or all that she does.[37]

Long story short, Wolf argues that the meaning we find in life is that we can love something or someone. One can love one's job and one's children, one's hometown and Beethoven's Ninth Symphony, just as one can love playing the piano or jogging. However, Wolf says, one can make mistakes in loving because one loves something or someone objectively unlovable. Whoever loves to treat other people cruelly commits a (moral) mistake, and each of us has certainly found out more than once in life that we have loved something or someone that really does not deserve our love (for one reason or another).

Here, it is not a deception that there can be errors in matters of love and preferences without there being only one object, activity or person worthy of our love. If one can be mistaken about a matter, it is objective at least in this respect, which does not mean the value had by a thing, an activity or a person one loves is therefore the same for everyone.

Subjective and objective are not necessarily mutually exclusive. Here, one can draw on a helpful distinction by John R. Searle, who has pointed out that something can be objective and subjective at the same time. There are facts that exist only because subjective states are involved in them (feelings, opinions, sensations, etc.). These facts include, for example, toothaches, feelings of love and thirst. These are **ontically subjective**. Nevertheless, they can be studied objectively, for instance psychologically, sociologically and physiologically, or we can simply have mundane opinions about them which can be true or false. Ontically subjective facts can therefore be **epistemically objective**.[38] Objectivity is not a neutral view from nowhere but an attitude towards a state of affairs, thanks to which one can be right or wrong. These distinctions are important in order to understand Wolf's thesis that there are objective answers to the question of whether something we love really deserves to be valued by us in an intense way. Cruelty and ruthless egoism may be loved by some people (Vladimir Putin comes to mind once more), but these behaviors do not objectively deserve our love.

Love is central to human life. As is well known, none of us passes infancy without having had some experience of love. One must have been loved in some measure and have loved back if one has come so far as to read these lines. What one experiences in love and thanks to it is the meaning *in* life, according to Wolf's basic thesis, with which one should agree. Of course, this insight is not new, which would indeed be surprising. If it is true that the meaning *in* life is to love and to be loved, this has been true ever since there have been human beings and other living beings who can experience and give love.

Love is linked to devotion. We gladly give ourselves to what we love. It does not bore us and it puts everything that happens in a favorable light, like the proverbial rose-tinted glasses one wears when one is freshly in love. Nevertheless, love is not immune to error, which we all also know from an external perspective when one of our friends has once again fallen in love with the wrong person.

Meaning in *life does not exist independently of being experienced. In this respect, it is* subjective. *Nevertheless, it can be evaluated by an independent person (be it a friend, a psychologist or a sociologist) from a third-person point of view, so the experience of meaning has an* objective *component. This objective component is closely related to ethics, that is, to moral values, which also exist independently of our subjective opinions.*

Because loving has an objective component interwoven with ethics, many people believe that meaning *in* life is related to moral devotion and commitment. This is partly true. The life of a saint certainly seems more meaningful to us than the life of a mass murderer, and rightly so. A person's preferences are always also ethically evaluable, although many of our preferences (such as the vast majority of sexual preferences) are morally neutral.

However, it would be an absurdly exorbitant demand if we were to identify the meaning *in* life with always doing the morally right thing, which Wolf already elaborated in the famous essay "Moral Saints" in 1982.[39] One does not live meaningfully only if one emulates Jesus, Mother Teresa, Krishna, the Prophet Mohammed or Buddha, whom we appreciate because they made it the goal of their lives to love humanity as such and to realize the good (be it with or without the help of the monotheistic God).

Wolf's approach of looking for meaning *in* life rather than drawing an all-encompassing meaning of life from a transcendent perspective has the advantage of holding that all people are capable of finding meaning in life *in* loving, regardless of a specific religious or ideological view. This is not just a subjective stance, but it tells us something about objective standards binding all people together.

The Meaning of Life is Not Nonsense

The attractive liberal pluralism associated with the concept of meaning *in* life must nonetheless not lead us to throw the baby out with the bathwater by abandoning the question of *the* meaning of life altogether. For the question of *the* meaning of life is by no means metaphysical nonsense, as held by tradition of linguistic-philosophical positivism which was influential in the last century. This line of thought claims value judgments are not about reality. The idea is that language can meaningfully speak only about those things and processes that we can investigate scientifically or empirically (e.g. by means of statistics and studies).

Since the end of the nineteenth century, the term "positivism" has been used to describe the view that much or even most of what our ancestors were passionately interested in (the immortality of the soul, the existence and will of God or the gods) was ultimately nonsense. Meaningful thinking could only be done about whatever could be measured by means of natural scientific methods. All other forms of knowledge are to be judged according to their contribution to the interpretation and justification of the scientific knowledge of nature. Logic and mathematics could just barely be saved, because without them there would be no scientific theories and thus no interpretations of measurement results would be conceivable. But ethics and aesthetics, not to mention metaphysics in the sense of a science of non-sensual or super-sensual objects, were suddenly considered nonsense after many millennia of human history.

Paradigmatic for this positivist stance is a famous essay by the philosopher Rudolf Carnap, who was to become one of the most academically influential philosophers of the twentieth century. In his 1931 essay "Overcoming Metaphysics through Logical Analysis of Language," he attempts to show that there are clear, logically and linguistically determinable criteria of meaning that would be violated by statements about such far-reaching questions as the meaning of life, the soul, God, etc.[40]

Even if the theses of this essay are no longer considered viable today, the suspicion that the question of the meaning of life is somehow nonsense, often repeated by Ludwig Wittgenstein, remained in the room. Thus Wittgenstein states in a series of seemingly paradoxical statements in his *Tractatus Logico-Philosophicus*:

> We feel that even if all possible scientific questions be answered, the problems of life have still not been touched at all. Of course there is then no question left, and just this is the answer.
> The solution of the problem of life is seen in the vanishing of this problem.
> (Is not this the reason why men to whom after long doubting the sense of life became clear, could not then say wherein this sense consisted?)
> There is indeed the inexpressible. This shows itself; it is the mystical.[41]

Directly after these remarks Wittgenstein claims the method of philosophy consists in uttering only "propositions of natural science," "i.e., something that has nothing to do with philosophy,"[42] which is an obvious contradiction to what he does, which is why he himself describes his own propositions "as senseless."[43] For the thesis that one should reduce meaningful language to propositions of natural science (whatever these may actually consist of) is itself not a proposition of natural science and thus not meaningful by its own lights.

Positivism failed time and again because of such self-contradictions. A self-application consists in referring a statement to itself. A simple example: whoever thinks all knowledge is actually only an expression of subjective opinions ("Everything is relative") expresses with this thesis something that cannot be only a subjective opinion. The positivists now thought the meaning of words and sentences could be traced back to a method of verification. The word "Cat" means what it means because someone at some point saw a cat and referred to it as "cat." And the sentence "It is raining in London" gets its meaning from the fact that you can go to London and see it is raining there. But if one infers that linguistic meaning generally consists in expressing empirically verifiable facts, one has with this very statement expressed an empirically unverifiable fact.

If the positivists of the so-called Vienna Circle, to which both Carnap and (marginally) Wittgenstein belonged, claimed that propositions about non-empirical things (such as God) are without any meaning because they cannot be proved by empirical studies, then this very statement that one cannot talk about God empirically already itself contains a contradiction under self-application. For the word "God" occurs in the statement. And this refers to something that itself cannot be empirically checked and verified. An apparently simple contradiction under self-application. Many positivists tried to prohibit self-applications. Thus logical and linguistic-philosophical arguments were put forward to show that self-applications always lead to paradoxes. But this did not stop the logically and mathematically most gifted minds of the Vienna Circle, among them the great mathematician Kurt Gödel, from making self-application the motor of ingenious mathematical proofs. So positivism gradually lost its plausibility, which was already only apparent.[44]

In any case, the positivist critique of metaphysics is not a sufficient reason for declaring the question of *the* meaning of life to be nonsensical and to make do instead with the more earthly search for meaning *in* life. For the positivist critique of metaphysics is considered to have failed according to today's philosophical standards, so that the question of the meaning of life, which every human being asks themselves at some point, is in any case no longer hindered by the reservation that the question is nonsensical or empty of meaning in some other way.

The great questions of humanity, which have occupied us since time immemorial and which almost every human being encounters at some point in his or her life, cannot be overcome by suppressing them and preferring to devote oneself to logical-mathematical-scientific research. This is an extremely meaningful and enriching occupation, in which many people find a meaning *in* life. But one must not conclude from this that the great metaphysical questions of humanity have somehow lost their importance. For this does not correspond to the facts and to human experience, and it is accordingly wrong and should not be defended under the banner of science, which is after all tied to the search for truth.

Nonsense is Sense-Deprivation

We must go a little further before we can approach the meaning of life safely, so to speak. For this purpose, we have to overcome the misguided view that reality, and thus also our occurrence in it, is meaningless in itself and at best receives value and meaning from the fact we give meaning to life. For how should something receive real meaning by our giving it meaning? This basic assumption of modern nihilism looks attractive only at first sight.

Rather, it is the other way around: our life seems to make sense to us, and we recognize this precisely when the meaning slips between our fingers in a crisis and we have the feeling of standing in front of nothing. What we then experience is nonsense *par excellence*, the feeling that ultimately absolutely nothing has any deeper meaning, that we are all just a ridiculously short blink of the eye in the universe, or "Darwinian mutants," as the narrator in Scarlett Thomas's novel *The End of Mr Y*, inspired by her reading of Heidegger, calls it.[45]

The meaning of life is not nonsense, but, rather, the absence of meaning is. Nonsense is sense-deprivation, and thus presupposes the experience of meaning. The experience of a deep meaninglessness befalls us when the relationships in which we find ourselves with other people, and ultimately with nature and society as a whole, change abruptly and we lose the thread of our life. Then we desperately seek to reconnect somehow with our lives, which does not always succeed, and not for everyone.

I am writing these lines in a hotel in Montreal during a lecture tour. The Canadian border has just been reopened, while in Germany an unprecedented Covid wave is still raging. The sudden change of place from an irritated and frightened German society to a different one, as well as the impression of a ghostly airport where basically no one arrives (word has not yet spread that you can travel to Canada again), triggered a feeling in me that my meaning intentions had faltered. I got into conversation with a homeless man on a walk who told me he had lost his job and his family because of mental illness. His life no longer had any meaning, and since then he has survived in constant despair, without seeing a way out. He told me he found life is meaningful in itself, but one should never lose the thread. It is difficult to find it again.

Unfortunately, such fates are a dime a dozen. Now think about all the experiences billions of people are having at this very moment and you will realize life is both full of meaning and full of nonsense, which opens up like an abyss when we lose our individual sense of meaning.

This consideration shall show that meaning is the starting point of every experience of its absence (meaninglessness) and not vice versa. We do not find ourselves embroiled in nonsense that we have to gloss over, as it were, so that our life seems to make sense to us temporarily. It's not by chance that we start our lives as infants with stories, games and daydreams we find meaningful. This corresponds exactly to the reality of human experience. Human beings are born dreamers. Meaning is expressed in our dreams and fantasies. If we are robbed of our dreams or lose them for some reason, without being able to find them again, nothingness and nonsense overtakes us.

The nonsense of nothingness can be so overpowering that it is not possible with one's own power to counter it with a meaning. Here "nothingness" is a name for the ultimate withdrawal of meaning. In order not to oppose the plurality of forms of life in liberal democracy against the human need to fathom the meaning of life, it is necessary to reverse the modern direction of explanation and to seek connection to the wisdom of humanity, which since time immemorial has assumed that we bear witness to an event of meaning, a history that is not just a hallucination of our brain useful for survival. Meaning is primordial and nonsense is an experience of sense-deprivation and loss. Just as life is prior to death, meaning is prior to meaninglessness. Life does not receive its meaning from death, nor is meaninglessness the ground of meaning.

Limits of Liberal Pluralism?

As the literary scholar Terry Eagleton points out in his now classic booklet *The Meaning of Life*, the question of the meaning of life puts liberal pluralism in its place by showing that certain metaphysical commitments (for example, to the unconditionality and inviolability of human dignity) are unavoidable if one actually wants to live in a liberal society.[46] It is true that in a modern democratic constitutional state everyone can find their own meaning *in* life within the framework of certain rules. That is, they can live out their preferences. Nevertheless, not everything is allowed even in a democratic constitutional state. The state restricts the unbridled freedom of citizens in many ways. At the same time, we have become accustomed to the fact that restrictions of freedom in the modern age are no longer justified with reference to transcendent, higher sources of meaning such as God and the gods. Yet these have been functionally replaced in the meantime by a problematic faith in science and technology. The latter justifies restrictions of freedom by referring to scientific imperatives (which also include economic 'necessities' provided by research institutes for policy advice). The faith in science and technology therefore merely seems to be ideologically neutral and not to touch the meaning of life. But appearances are deceptive.

We live today in most industrial societies of the formerly so-called West with the rather implicit idea that the meaning of life is to do everything possible so that scientific-technological progress leads to economic growth. From this in turn we expect an improvement of our living conditions. And for this fallacy we are rightly criticized by those who suffer most from it, because we have destroyed their life-worlds through exploitation of natural resources, imperialism, colonization, and other forms of hubris.

Why should it actually be a sensible human goal to do everything in our power to develop new technologies at an ever faster pace in order to create even more opportunities for consumption and to live even faster? If one considers this to be sensible, one recognizes precisely thereby that one has become entangled in an ideology that is anything but ideologically neutral.

The unconditional belief in scientific and technological progress today hides the fact that this progress achieves the opposite of its goal. While we know how close we are to self-inflicted extinction, this very extinction is driven by a worldview that measures all social events by parameters of economic growth – parameters closely intertwined with our belief in science and technology as a form of earthly salvation.

This misconception is embodied by the team around the fictional US president Janie Orlean (played by Meryl Streep) in Adam McKay's 2021 film *Don't Look Up*. The plot revolves around a small group of astronomers who discover that a gigantic asteroid is hurtling towards Earth and will cause mass extinction in a few months. The populist president then teams up with the head of a futuristic tech company, tech guru Peter Isherwell (played by Mark Rylance), who wants to avert the catastrophe by blasting the asteroid into small pieces so that the raw materials released in the process could also be used by the US economy. The plan fails because some of the explosives used don't work as well as the tech guru promised. Spoiler: humanity fails to stop the catastrophe. The reason for this depends ultimately on the interpretation of the film. One surely legitimate interpretation blames the failure on the connection between the White House and big tech, which is based on a solutionist idea of salvation to which the Belarusian technology theorist Evgeny Morozov has dedicated a relevant book.[47] **Solutionism** is in this context the idea that technology can solve any problem; if necessary, we just have to invest even more vast sums in it. The fictitious president, however, should not be understood as an allusion to Donald Trump, as many interpreters thought at first glance; she embodies not simply an anti-scientific stance of "alternative facts" (another word for bullshit) but also the unconditional belief in the technological controllability and economic exploitability of all resources. For this reason, unlike Trump, she does not criticize the American intellectual cadre but mistrusts the discoverer of the asteroid only because he teaches at a more or less insignificant university.

If the idea of a completely metaphysics-free society is incompatible with liberal pluralism, it does not follow that a certain religious metaphysics should shape our institutions. Quite the contrary. Liberal pluralism, of course, rejects above all the idea that the state is organized theocratically, i.e. that representatives of a particular religion determine the fundamental course and institutions that affect all of us as citizens. Nowadays, at least to a large extent, we keep the will of God or the gods out of state affairs so as not to endanger the creative freedoms of individuals. But, in doing so, we interfere with the creative freedoms of all those who do not agree with liberal pluralism. If one considers the whole of humanity, the group of those who consider liberal pluralism an unacceptable worldview is certainly not a small minority. It was precisely the great illusion of the 1990s in the West to believe that after the fall of the Berlin Wall liberal pluralism and the democratic rule of law would displace all other political systems – theocracies, authoritarian systems, dictatorships, and so on. On the contrary, history is once again all too open, as the current geopolitical crisis situation impressively proves. This, of course, does not refute Francis Fukuyama. His thesis of the "end of history" had a normative meaning. He wanted to say that liberal democracy is based on a foundation of values that can no longer be surpassed and that it thus also has prospects of prevailing in world history because it is interested in facts and (unlike Putin's paranoid metaphysics of history, say) in truth and reality.

In any case, liberal pluralism is not an ethically neutral starting point, thanks to which everyone can be happy according to his or her own liking, but the expression of a value judgment based on the fact that we should allow as many people as possible as much individual freedom as possible to find their ideas of a successful life and thus their meaning *in* life. Alternative value concepts of nationalistic ideologies, on which autocracies and dictatorships feed, rely on local cults and specific religions to oppose the universalism of enlightenment and prevent all people from coming together. The value foundations of contemporary Afghanistan or Putin's kleptocracy combat the notion that everyone is entitled to find meaning in life. This notion is expressed in a liberal attitude towards people who think differently. But of course the tolerance of liberal democracy also has its limits, because democracy is entitled to adhere to its values and to defend them if necessary.

I consider liberal pluralism to be ethically recommendable in principle provided it leads to moral progress. Liberal pluralism aims at moral progress, because it is concerned with allowing, protecting and promoting a variety of legitimate ways of life without prior metaphysical restrictions.

The justification of liberal pluralism should not pretend it is ideologically neutral and has no opinion on the meaning of life. Whoever thinks they want to be secularly modest or agnostically reserved with regard to the question of the meaning of life has thereby already taken an ideological or metaphysical position

— which one can easily allow oneself in a democratic constitutional state that also allows this agnosticism. One cannot be neutral towards life and leave the creation of meaning to individuals without thereby setting an institutional and collective course, and thus ultimately affecting all people, giving an implicit or explicit answer to the question of the meaning of life.

This insight is at the heart of the political theory of the Greek-French philosopher and psychoanalyst Cornelius Castoriadis. In his main work *The Imaginary Institution of Society* he comes to the conclusion:

> Every society up to now has attempted to give an answer to a few fundamental questions: Who are we as a collectivity? What are we for one another? Where and in what are we? What do we want; what do we desire; what are we lacking? Society must define its "identity," its articulation, the world, its relations to the world and to the objects it contains, its needs and its desires. Without the "answer" to these "questions," without these "definitions," there can be no human world, no society, no culture — for everything would be an undifferentiated chaos. …
>
> Of course, when we speak of "questions," "answers," and "definitions," we are speaking metaphorically. These are not questions and answers that are posed explicitly, and the definitions are not ones given in language. The questions are not even raised prior to the answers. Society constitutes itself by producing a *de facto* answer to these questions in its life, in its activity. It is in the doing of each collectivity that the answer to these questions appears as an embodied meaning; this social doing allows itself to be understood only as a reply to the questions that it implicitly poses itself.[48]

Thus we owe Castoriadis an important insight: society does not exist without our imagining it — each in a slightly different way. It is "imaginary," as Castoriadis says. Our ideas about society do not have to be correct in order to be effective. It is also part of society that we imagine it one-sidedly and wrongly, that we are in error about it in one way or another. A society of omniscience is unthinkable; it would at any rate no longer be human. Here it transpires once more that we are in a paradigmatic way a part of nature: We are finite and vulnerable, which we only partially disguise in our social life.

Liberal pluralism is embedded in a conception of society whose tolerance is limited. Every society excludes something internally and externally. Liberal pluralism excludes illiberal democracy as well as autocracy and does not allow any religion to become the state religion. Liberal pluralism finds it particularly difficult to recognize that it itself rests on an ideological foundation and, consequently, on an implicit answer to the question of the meaning of life, because it has too long assumed that moral relativism is a necessary condition of liberal democracy. But just the opposite is the case. Moral relativism, which asserts there are different moralities (Western as opposed to Chinese, for example), is not compatible with liberal democracy. It is anti-enlightenment and illiberal. In the end it even behaves tolerantly towards the *de facto* enemies of freedom and peace.

Moral relativism delegates the stability of value settings and thus its own implicit setting of a meaning of life to a purely bureaucratically organized society, which is not supposed to have a strong foundation of values. The fact that liberal democracy does in fact have a strong value foundation, which is not an arbitrary setting among many other possible ones, speaks against the idea of a technocracy or expertocracy in which political decisions are seen never as questions of value but as something that can be seamlessly derived from data, studies and scientific knowledge. Life is not altogether a scientific-technical problem. How to educate one's children, what one believes in, which art forms one appreciates, what one stands for even to the point of sacrificing one's life if necessary – none of this can be determined in a scientific way. It is rather the case that people who dedicate their lives to scientific research (which is unquestionably a worthy pastime, as one never gets bored in view of the almost infinite complexity of the universe) do this because they see a meaning in it. The beauty of mathematical equations, the surprises that lurk in all corners and ends of the universe, show that reality is full of structures at which one can only marvel.

The limits of liberal pluralism lie precisely where we have to acknowledge that we cannot avoid conceiving of life as something meaningful, which brings us to the question, as yet unanswered, of how deep the meaning of life actually goes. We do not know enough about life to exclude that a creator expresses him- or herself in it. Of course, we cannot prove the existence of a creator either, because it is always conceivable that the formation of structure in nature is self-sufficient and can be explained from its own resources. We will also never solve the riddle of life as long as we are not at the same time able to answer the question to what extent the mind actually belongs to nature and to what extent we ultimately do not completely fit into nature, despite or even because of our animality. In some way we do not fully understand, we rise above the universe (in the sense of the totality of what can be investigated by natural science) without being able to renounce it in this life. Liberal societies not only rely on being indirectly metaphysical and evaluative by preferring a secular legal order that takes account of ideological plurality. In addition, they need – at least so far – a basis, which today is seen above all in the coupling of natural science, technology and economic growth. This too is a metaphysics, an expression of a worldview that has been showing its fatal shortcoming in our age of existential risks and thus of potential self-extinction.

Because value orientation is unavoidable and cannot be circumvented in the name of justified liberal pluralism (since this in turn is based on strong valuations), it is incoherent to delude ourselves into thinking we could as it were pursue value-free politics by following science and technically exploiting its 'results.' In any case, such a politicized view of science proves on closer examination to be neither value-neutral nor recommendable, as it achieves even the opposite of what it aims at, namely the undermining of our basis of survival on the only planet on which we will ever succeed in leading a meaningful life.

Who We Are and Who We Want to Be – Radical Autonomy and the New Enlightenment

The experience of meaning is the starting point of all our thinking and acting. What we do with commitment and interest, we experience as meaningful. If this experience of meaning diminishes, it is not possible for us to continue doing what we are doing, unless we are forced by external circumstances to do alienating work – which to be sure is true for the vast majority of people living today.

This is reason enough to design a political utopia based on the idea that it is a human right to be able to lead a life that enables and strengthens experiences of meaning. We cannot only grow *quantitatively*, e.g. by increasing our wealth, thereby further promoting the affluent society under which our planet and the vast majority of people suffer who do not belong to the lucky few born in the right place or into the right class. We can and should all also grow *qualitatively*. That means we should be capable of pursuing long-term goals of self-education.

From this can be derived the basis of a New Enlightenment, which has been called for by various thinkers worldwide for some years now. This New Enlightenment can draw on the idea of radical autonomy, which – contrary to what some might now expect – is not an atomistic model of society made up of individuals, because autonomy can only be achieved under social conditions.

Let us again proceed step by step. In general, **autonomy** (from Greek *autos* = self and *nomos* = law) is the capacity of a living being, paradigmatically a human being, to give itself a law, i.e. to measure its own actions intentionally against a standard which one must oneself recognize in order for it to apply. This is distinct from the notion of being guided in one's self-image by something else (whether divine commandments, superiors or other authorities). People who act autonomously do what they do because they are guided by rules they have given themselves. These rules may include informing oneself and trusting the judgment of others for good reasons, which is a different attitude than blindly following some authority.

With the idea of autonomy, one can now go one step further. One can make the thought of autonomy itself the basis of acting, which presupposes that we consider also the autonomy of others. For others are also capable of doing something in the light of an idea of what they think is right. Those who base their actions on the idea of autonomy are thus not limited to themselves, for autonomy does not consist in elevating oneself to authority. Instead of autonomy, that would be an internalized cult of personality, i.e. narcissism.

Every person has an idea of who they are, who they were and who they will be. We compare this idea daily with who we want to be. Being someone consists in giving an implicit or explicit answer to the question of who we are and who we want to be. This is an expression of autonomy.

Now this applies to every person. The same ability of being a human being thus ultimately radiates from every person, an ability we merely interpret and exercise differently. This means we can orient ourselves by the ability of being human when we ask ourselves who we want to be. We also express this by praising someone as particularly "humane." As we have already seen in Part I of the book, the concept of the human being is not only a descriptive concept – designating a biological species – but at the same time a value concept.

Of course it is true and should not be concealed that the image of the human being [*Mensch*] was traditionally oriented towards men [*Mann*], which is reflected in the etymology of the German word (as also holds for other languages, e.g. the English *man*, the French *homme*, etc., but not, for example, the Chinese 人, pronounced rén, which is an ideogram, i.e. in this case an image for a general human figure with two legs). However, nowadays it is fortunately not difficult to become aware of this gender discrepancy and to throw the completely absurd idea on the garbage heap of history: the idea that men (in whatever 'manhood' may consist) are human beings in a more important way. Human beings exist in all genders and in every range of sexual self-determination. Thanks to several phases of sexual emancipation since the nineteenth century, social institutions have at some point come to realize we are autonomous in our sexuality – although today there is sometimes bitter controversy about how far sexual self-determination should go (another important topic beyond our scope here).

In order to understand more precisely the structure of radical autonomy, which underlies the demand for a New Enlightenment, it is worth making a small digression into the thought of Immanuel Kant and Georg Wilhelm Friedrich Hegel, who should be brought together at this point. What I have called self-determination so far, Kant and Hegel call "will." Kant remarks on this in an important passage of his *Groundwork of the Metaphysics of Morals*:

> Everything in nature works in accordance with laws. Only a rational being has the capacity to act *in accordance with the representation* of laws, that is, in accordance with principles, or has a *will*. Since *reason* is required for the derivation of actions from laws, the will is nothing other than practical reason.[49]

To act according to principles here means only that we are able to imagine who we are and who we want to be and to derive maxims (as Kant calls them), i.e. specific ideas concerning our courses of action, which at any given time are connected with a more or less well-thought-out way of life and (for some people) with a whole life plan.[50]

A human life extends in time. We not only live in moments but embed them in some way or other in a larger temporal context, so that we probably all work on our life stories. The socially shared life reality of people in a community is therefore historical, an unfolding of stories.[51]

What Kant regards as acting *"in accordance with the representation* of laws" is the basic form of autonomy. The autonomy of agents can be restricted and extended, depending on how they are embedded in larger social formations. As individuals human beings are conceptually independent from others. As such, we have autonomy to lead our life as we imagine it. If there were no others (including non-human beings) to whom we automatically have moral obligations, we would be free to do as we please without restriction. As soon as what we do has consequences for others, our autonomy is changed, because it reaches its limit where it touches the autonomy of others. At this point, the scope of our agency is not only restricted by the presence of others but also expanded, as most things we do are such that we can only do them with others. Our individual freedom thereby becomes social freedom which incorporates the autonomy of others.

And it is precisely this consideration that leads Kant, and Hegel in his wake, to the realization that law assumes the function of establishing social norms that provide information about which restrictions on autonomy are legitimate at a given historical point in time. Political freedom therefore consists paradoxically in first restricting our original individual freedom and second expanding it by disclosing a broader space for joint action. This restriction does not automatically take away something we would like to have but enables social freedom, without which we could not do most of what is meaningful for us. Our original freedom is therefore the empty idea that we could actually do everything if others did not exist – except we could then do almost nothing at all of what we actually prefer to do!

And this is exactly our radical autonomy: we are able to limit the scope of our individual freedom in order to expand it, because individual freedom can only be realized socially.

To be sure, the details of the restriction of our freedom (and thus our autonomy) require justification. Restrictions must not only be proportionate in the legal sense but must also correspond to moral facts and thus values. The democratic state under the rule of law must therefore be prepared to examine its procedures by recourse to soul-searching and other methods of ethics, which is why we expect judges, politicians and other decision-makers to demonstrate moral standards by explicitly justifying their actions when necessary.

So far, so good. But this consideration threatens to introduce another dualism, i.e. a tension between two poles we often encounter in public debate under the keywords of freedom vs. security. If freedom originally means autonomy, and if the role of others is defined by their justified claims to restrict my autonomy, society ultimately appears as a gigantic restriction of freedom. But without inner and outer security no freedom is possible and without freedom no inner and outer security. Society must not, however, become a burden for the autonomous subject – an idea is at the heart of the existentialism of Sartre, for whom other people are on the one hand indispensable but on the other primarily troublesome obstacles. This corresponds to Kant's familiar phrase about the "unsociable sociability" of the human being.[52]

However, as stated, this consideration overlooks the fact that most of what we want to do is only possible because we do it with others and for others. For this reason, other people not only limit our autonomy but also promote and extend it by creating social facts together. Those who want to play soccer alone miss out on the actual experience of playing soccer. But as soon as other people are present (even if it is only the players and the referee), the autonomy of the individual players is limited – if only because every player wants to score a goal, but the purpose of the presence of the others is precisely to prevent this if possible. Therefore, it is wrong to understand autonomy and society or individuality and sociality as somehow incompatible.

And this is where Hegel comes into play. It is not enough to understand our autonomy as a free will standing over and against society and its constraints. For then we cannot understand how social freedom is supposed to be possible. But we know there is social freedom, because the scope of action for people who can fall back on a functioning health and education system, or even (unfortunately unthinkable in Germany for a few years) on a functioning infrastructure (punctual trains!), is wider than the scope of action for people who live in impoverished dictatorships such as North Korea.

Social freedom consists in expanding our autonomy through cooperation with the autonomy of others. Society is therefore not the Sartrean hell on earth in which everyone mutually restricts one another but an organization whose purpose is the expansion of our scope of action through a morally justified restriction of individual freedom, which we accept voluntarily in order thereby to achieve more social freedom.

In the introduction to his *Elements of the Philosophy of Right*, Hegel shows in a still exemplary way that the will is not a pre-social, asocial or even anti-social faculty. Rather, according to Hegel, "the Idea of the will ... is *the free will which wills the free will*."[53] According to this view, the law (our rights) is the realization or, as Hegel says, "the existence" of this idea of social freedom.

So we come again to the project of a New Enlightenment. This assumes we can and should work together not only to demand autonomy in the form of rights, and to impose it on ourselves and others as duties, but that we can turn autonomy itself into the norm of our actions, what Hegel calls "self-conscious freedom."[54] We can understand this as radical autonomy.

Radical autonomy consists in recognizing the ability of people to lead their lives in the light of an idea of who we are and who we want to be as a source of value.

We thereby always measure our individual autonomy against the autonomy of others. Thereby, we understand that there is social freedom, and that freedom is accordingly not opposed to socialization.

This insight cannot, however, be meaningfully articulated without accepting that there is a reality that reveals itself in the structure of radical autonomy and social freedom. By this I mean that human sociality does not exist only in our individual heads as a neural representation or as some other kind of imagination. Society is a reality, and it does not take place only in our heads. Hegel is therefore correct when he describes social freedom as "objective mind," i.e. as the realization of radical autonomy on the part of rational human beings under historically variable conditions. Society is therefore not a spectacle of nature. We do not completely belong to nature but are all parts of a historically unfolding objective mind that cannot be reduced to the sum total of our individual representations of it. What people do by creating institutions of social freedom (from children's games to the rule of law) differs in principle from the emergence of complex structures from simple natural units. For institutions also consist of intentions; they are an expression of what is called "collective intentionality" in social philosophy, which means the partly intentional, partly unintentional interaction of people. Institutions cannot be reduced to swarm intelligence; a supreme court, say, is not even remotely akin to the organization of a state of insects.

A society is not a whirlwind that results from the interplay of elements and forces in a physically describable field. Society arises from the fact that we compare our attitudes with one another and form an explicit, symbolically (linguistically, in writing, artistically, medially, etc.) encoded image of what we and what others think, feel, and so on.

On this basis, the New Enlightenment demands that we engage in a spirit of trust that strives for cooperation across the different sectors of society with the clearly set goal of achieving moral progress and thus more social freedom together. In this way, we can overcome the fatal and absurd idea that capitalism is an ultimately destructive system of greed for profit and exploitation of people and nature, but that it is also inevitable and has no alternative. For there is no such thing as pure capitalism, as many critics of capitalism imagine. Rather, since the nineteenth century in Europe, capitalism has been increasingly limited by the simple socialist (and correct!) idea that a society, and thus an economy, cannot function in the long term if it is primarily controlled by greed for profit, economic growth and exploitation – an idea that, incidentally, Adam Smith had already rejected.[55] Predatory capitalism, which is based on the idea of competition for scarce commodities and not on the cooperation of social animals in need of mutual aid, is not inevitable. The idea that any form of predatory, competitive economic system is somehow baked into our nature as animals would be at best social Darwinism. That is: the mistaken projection of biological concepts onto human social behavior.

At this point, the concept of life comes back in. Based on the research of the famous biologist Lynn Margulis, in recent years the idea is becoming more widespread that cooperation is more fundamental for the evolution of life forms than competition.[56] Symbiosis and mutualism, i.e. the exchange of positive abilities between elementary life forms (such as bacteria and algae cooperating with each other in lichens), are the normal case – and not the fierce competition for scarce survival goods. Based on microbiological findings, which show, for example, that bacteria have been evolutionarily fused into our cells such that we would not even exist without a very close connection between different life forms, some biologists have come to the conclusion that we are not individuals but complex webs of different living beings. This is another way to overcome the concept of 'animal' and to think of life in a completely different way than neo-Darwinism and social Darwinism have in mind.

Here we can once again turn to the physicist and complexity theorist Dirk Brockmann, who derives a social utopia from the work of Margulis and other facts the life sciences have discovered in recent decades:

> For 100 years, neo-Darwinism and social Darwinism have cross-fertilized and led to fatal concepts of life and economy: unbridled growth, monopolistic corporations, uniformity and loss of diversity. Perhaps it is time to learn from nature's most successful strategy and adopt it in societal and social structures: cooperation.[57]

The fundamental role of cooperation is of course taken into account in socio-economic reality, as in many states society and the economy are by no means organized as the social Darwinists described. Apart from models such as the social market economy or the welfare state, which take into account the concept of social freedom at the level of markets as well as the behavior of market participants, we are unquestionably in the twenty-first century in the age of a major economic transformation. For, if we want to survive, we must succeed in ending the age of fossil fuels and in inhabiting the planet together in a way never seen before. At this point, Brockmann's reference to the biology of cooperation is a step further, since he shows humans could not exist as animals without already being dependent on cooperation at the micro-level of cellular organization. Cooperation runs all the way down to the cellular level, and we could not even survive without it.

Social Freedom and the Meaning of Life

We must trust ourselves again to develop utopias of social freedom, thanks to which we understand society as an organization that makes it possible to ensure moral progress in the long term. Such progress is fragile and is threatened again and again – which we are currently experiencing in the form of rapidly rising right-wing populism (which in reality is a rather dull national egoism), among other phenomena.

The objective mind of humanity, expressed in our institutions, is realized social freedom. Thus, the modern democratic constitutional state has a value foundation that goes far beyond individuals finding meaning in their lives. Our institutions have a history that is interwoven with art, religion, philosophy, science and technology. Because of this history, value concepts and therewith answers to the question of the meaning of life have dissolved into our institutions. It therefore ultimately hits the mark that "The rise of the state [is to be understood] as a process of secularization," as the title of a notorious essay by the constitutional lawyer Ernst-Wolfgang Böckenförde puts it. In this essay he articulates what has become known in the German debate as the so-called **Böckenförde dictum**, which states: "The liberal, secularized state is nourished by presuppositions that it cannot itself guarantee."[58]

Böckenförde, in the wake of Carl Schmitt, assumes that value judgments concerning the meaning of life (represented in a paradigmatic way by religion) are prerequisites of the liberal state but cannot be guaranteed by it. But the opposition of institution and strong value judgment presupposes that the institutions are secularized, which is not true, as they make judgments in the context of an unconditional obligation to human dignity, which they can and should guarantee. The state is not a detached social stratum which would be independent of the value judgment of individuals.

Respecting human dignity means recognizing in other people human beings to whom we have unconditional, non-negotiable obligations. One of the reasons that human dignity is recognizable to us is that we are already in the medium of meaning when we ask ourselves what to do in the face of a decision-making situation. In the medium of meaning, we recognize that others also experience meaning, that they are subjects who (for all their differences) are fundamentally similar to us. One can consider it a mandate of the modern constitutional state to allow as many people as possible to search for the meaning of life, so that together we can better discern how to further expand our social freedom using the guide of the idea of moral progress.

This brings us to the heart of a widespread conception of the meaning of life. Many people think that the meaning of life consists in choosing the right way of life, which always presupposes that one is looking for the good as a guiding star. A path of life that is paved with nothing but morally reprehensible actions does not seem to us to be "worthy of happiness," to invoke once again one of Kant's thoughts.[59] Accordingly, a villainous dictator or torturer who leads a happy life is not supposed to exist. Kant defines happiness as "a rational being's consciousness of the agreeableness of life uninterruptedly accompanying his whole existence."[60]

Against this background, one can give an answer to the question concerning the meaning of life. According to this view, the meaning of life consists in attaining happiness by working together to be worthy of happiness. In concrete terms, this corresponds to the view, widespread in the world's religions, that the meaning of life consists in the fact that we are tested in this life of ours with regard to the morality of our actions. Accordingly, we should try to strive for a way of life that enables us and others to work together for the moral progress of humanity. We are to grow morally together in order to live morally respectable lives – guided by values – in economically functioning societies. Here, "morality" refers to what we should do or refrain from doing, because we are all human beings who by virtue of our humanity have mutual obligations and rights (whether these are recognized by just institutions or trampled upon by unjust ones).

Kant himself refers to the combination of (economically producible) happiness and (morally attainable) worthiness of happiness as "the highest good,"[61] which results from the fact that we are destined – whether we like it or not – to recognize what is morally right and to realize it in a socially prudent way.

Kant gets to the heart of the matter by explicitly considering radical autonomy, i.e. the "idea of the absolute worth of a mere will,"[62] as the destiny of humanity. Unlike other living beings, the human being is not only equipped with instincts to achieve its individual or family happiness but equipped also with reason, which requires us to be interested in the well-being of humanity regardless of our individual preference.

> In the natural constitution of an organized being, that is, one constituted purposively for life, we assume as a principle that there will be found in it no instrument for some end other than what is also most appropriate to that end and best adapted to it. Now in a being that has reason and a will, if the proper end of nature were its preservation, its welfare, in a word its happiness, then nature would have hit upon a very bad arrangement in selecting the reason of the creature to carry out this purpose.[63]

I interpret this passage as follows: organisms, in which eons of evolutionary selection are documented, are functionally ordered. The fact that we have eyes to see and a heart to maintain our blood circulation has to do with the fact that, as living beings, we are adapted to our ecological niche. If the meaning of our life were exhausted by the fact that we simply put ourselves in the safest possible position at every moment in order to secure ourselves and our offspring, it would not be possible to explain why we do so many things that put ourselves and others in danger.

What we do is often linked to rational representations of the consequences and meanings of our actions. For example, if you join a gym to stay healthy, you are investing in the future of your health. In order for gyms to exist, people have to invent equipment, sign contracts, construct buildings, and so on. All these activities by no means serve only the momentary well-being of those who carry them out; rather, they make sense only in the context of rather complex planning processes. These processes require reason in Kant's sense, i.e. the ability to think in larger contexts. If one examines a human society, one will quickly realize that most people are definitely not merely hedonists, only bent on having feelings of happiness in the moment. Rather, we are all rational in some way, planning for the long term. We even expect society to enable us to do this, which is why social freedom is essential to our activities. The purpose of our ability to reason must therefore lie in something other than just enabling us to live comfortably in the present moment.

The original Enlightenment view of the meaning of life maintains that the human being has a destiny, that it is in our nature that there is a purpose which we have to fulfill as human beings. The extreme nihilistic counter-thesis, that our life has neither meaning nor purpose, is in turn an answer to metaphysical questions. Those who believe that the meaning of life consists at best in gathering as many experiences of happiness as possible, or even in accumulating as much wealth as possible, overlook the fact that the form of our freedom is social. Whether we like it or not, humanity is ultimately a community of destiny, a fact that has become palpable for all of us in the age of the Covid pandemic and the ecological crisis that is certainly not going to disappear any time soon.

Let us hold then that the widespread view is on the right track that the meaning of life is revealed in the human mandate to create sustainable structures of social freedom that allow (as far as possible) every human being to spend a happy and happiness-worthy life in the community of human beings. Because we humans are rational and accordingly capable of realizing that on moral grounds we owe something both to one another and to other forms of life, it is our mission to realize the utopia of perennially increasing social freedom.

This option to give a direction to the question of the meaning of life is neutral with regard to the additional question as to whether our ethical cognition (our moral compass) is a sign of a higher power or whether moral claims follow from our form of life alone without harking back to a higher origin. For the good, the neutral and the evil are not morally loaded because a transcendent power originally gives them their normative force. The fact that one should not kill someone just because we do not like them (to choose an obvious example) is not due to a prescription by God or nature. The normative force of moral facts need not (and ultimately cannot) be derived from a source other than themselves. Moral reality is a reality of its own making.

Despite affinities with his moral philosophy, at this point we must correct Kant too insofar as he thought that reason is so radically autonomous that we can deduce what we should do or refrain from doing from claims of reason alone. On the contrary, moral claims, even if they are recognized by reason (but also by our feelings and socially acquired emotions), are not enacted by reason. Someone who saves a drowning child does not do so because reason dictates it but because they intuitively and correctly assess the moral significance of the situation without having to go through any specific process of reasoning.

The original Enlightenment and religious idea that the meaning of life is that we are destined in our lives to work together as a human community of destiny to make the best possible life feasible for all human beings, an idea which presupposes incessant moral progress, is therefore correct. Let us call this the Enlightenment answer to the question of the meaning of life.

As already mentioned, this Enlightenment idea is neither secular in the sense that the existence of God or the value of religion is denied, nor is it based on religion, because moral claims do not have to be justified by any authority that goes beyond what we owe to one another already because we are human beings. And it is certainly not specifically 'Western,' 'European,' or even 'Eurocentric' that there is a meaning of life which consists in the mandate of humanity to take responsibility for ourselves and for that over which we have power. Rather, this is a conception that has been widespread since time immemorial and that must be appropriated and brought to mind time and again.

Why Science Has Not Discovered That Life Has No Meaning

But is this primal Enlightenment vision of the good not naïve and obsolete? Today the following worldview is rather widespread, which suggests that there can be no meaning of life. According to this view, at the beginning of the universe (shortly after the Big Bang) there arose structures and forces which, after billions of years, led to the formation of planets. On at least one of these planets, Earth, life appeared at some point (it is not known exactly when and how). It then developed according to the laws of evolutionary theory and gave rise to all known forms of life. At the provisional end of this process we find the human being, who can visualize this process and realize that he is not at the center of any meaningful occurrence but only a side effect of the cosmic evolution of the total material-energetic system – stardust which has made it to short-term self-consciousness.

Indeed, one might believe that it would border on absurd hubris if humanity were to consider itself especially important in view of the cosmic vastness of which we are aware thanks to modern physics. Against this hubris, a suspicion of meaninglessness was expressed with particular prominence by Arthur Schopenhauer and Friedrich Nietzsche. Nietzsche formulates it thus at the beginning of his influential essay "On Truth and Lies in the Extra-Moral Sense":

> In some remote corner of the universe, poured out and glittering in innumerable solar systems, there once was a star on which clever animals invented knowledge. That was the haughtiest and most mendacious minute of "world history" – yet only a minute. After nature had drawn a few breaths the star grew cold, and the clever animals had to die. – One might invent such a fable and still not have illustrated sufficiently how wretched, how shadowy and flighty, how aimless and arbitrary, the human intellect appears in nature. There have been eternities when it did not exist; and when it is done for again, nothing will have happened. For this intellect has no further mission that would lead beyond human life. It is human, rather, and only its owner and producer gives it such importance, as if the world pivoted around it. But if we could communicate with the mosquito, then we would learn that it floats through the air with the same self-importance, feeling within itself the flying center of the world. There is nothing in nature so despicable or insignificant that it cannot immediately be blown up like a bag by a slight breath of this power of knowledge; and just as every porter wants an admirer, the proudest human being, the philosopher, thinks that he sees the eyes of the universe telescopically focused from all sides on his actions and thoughts.[64]

In this passage, Nietzsche compares our intellect with nature and comes to the conclusion that it is "purposeless." This result is justified among other reasons by the indication that it did not exist for a long time and at some point will no longer exist. But why should our intellect or our life be meaningless and purposeless just because we have come into being under natural conditions – which we can never completely survey and control in detail?

One can see that Nietzsche himself does not completely trust this explanation if one takes a closer look at the passage. For then one notices that his musings about nature and our position in it are explicitly based on a "fable." This fable, however, is certainly not a result of natural science but a comparison of a conception of the range of our intellect with categories of nature. Thus, Nietzsche expresses here a value judgment rather than objective knowledge that we would somehow be forced to accept as a matter of fact. Nietzsche wants life to have no meaning because he cannot recognize the meaning of life in the face of the anonymous forces of nature.

At the same time Nietzsche criticizes "the philosopher" in this passage but does not himself engage in anything other than philosophy when, unlike the Enlightenment thinkers, he concludes that life is purposeless because it comes about through natural processes. But great early modern natural scientists such as Newton or Leibniz, who, unlike Nietzsche, made groundbreaking contributions to the mathematics and physics of the universe, saw in the human ability to comprehend nature an indication that our intellect occupies a special position.

At this point, an insight from the American philosopher Thomas Nagel comes into play.[65] According to him, the idea that life, when compared with the great cosmic whole, has no meaning overlooks the fact that this very idea is the effect of a certain point of view. The fact that we may feel small in the face of cosmic vastness and that we know that everything we are and do will eventually disappear without a trace in no way proves the meaninglessness of life; it shows only that we cannot make sense of it from an external perspective on a universe that does not care about our needs.

In one of the most famous philosophical essays of the last fifty years, Nagel asked the now proverbial question: "What is it like to be a bat?"[66] He suggests that we can never know what it is like to be a bat because we have no access to the bat's subjective perspective, its experience. But whatever it may be like to explore one's environment by means of a sonar system and to hang from the roof of a cave, what is certain is that we can ask ourselves such questions and take an outside perspective, even on ourselves.

At this point of his reflection, Nagel rightly emphasizes that the world religions have ready an answer to the meaning of life precisely because they imagine that there is nothing beyond God.[67] We can see ourselves from the outside, i.e. as living beings which are at an ultimately random place in some galaxy cluster. But if God is an all-embracing reality, which assigns a specific meaning to everything, the meaning of the divine presence can no longer be questioned. If God is understood as a source of value, the absolute good, the suspicion of meaninglessness in relation to life naturally becomes superfluous. With God as a source of value and meaning, the question of meaning disappears, which is one of the points of the monotheistic theologies, which accordingly equated God with the good itself in some of their ancient varieties.

But you don't have to go that far to find the meaning of life. For why should one conclude from the comparison of some physical quantity (be it as vast at the entire cosmos) with the significance of our life that the latter is meaningless? The physical, biochemical and other realities that we can establish as facts on the basis of natural-scientific research teach us nothing at all about the meaning of life. It is not a research result of modern physics or biochemistry that life has no meaning and humanity has no purpose. Nagel writes:

> What we say to convey the absurdity of our lives often has to do with space or time: we are tiny specks in the infinite vastness of the universe; our lives are mere instants even on a geological time scale, let alone a cosmic one; we will all be dead any minute. But of course none of these evident facts can be what makes life absurd, if it is absurd. For suppose we lived for ever; would not a life that is absurd if it lasts seventy years be infinitely absurd if it lasted through eternity? And if our lives are absurd given our present size, why would they be any less absurd if we filled the universe (either because we were larger or because the universe was smaller)? Reflection on our minuteness and brevity appears to be intimately connected with the sense that life is meaningless; but it is not clear what the connection is.[68]

The impression that we are small in comparison to the universe and therefore meaningless is a value judgment and not a neutral statement concerning the size relations of humanity and the cosmos. No matter how the physically explorable universe is constituted, from its size, age and structure follows neither that life has a meaning nor that it has none. The question about the meaning of life is just (and here Wittgenstein got it right) not a scientific one.

Rather, it is an ethical question. Ethics is about what we humans should do or refrain from doing just insofar as we are humans. It is about what we should all do under certain conditions. Now, we human beings are those minded beings who live our lives in the light of an idea of who we are and who we want to be. Let us call an answer to the question of who we are and who we want to be a human self-portrait. Each of us has some conception of humanity that changes throughout our lives. Since, with reference to our human self-portrait, we see ourselves as something specific in one way or another (as an animal, as an immortal soul, as a neural pattern, as a reincarnation of a past life, etc.), we determine ourselves. This is what I call **existential self-determination**.

Thanks to our capacity for existential self-determination, we define a framework within which we make concrete ethical value judgments. Existential self-determination is accordingly never value-neutral: depending on what we consider ourselves to be in our humanity, we develop an individual and collective moral compass.

The unavoidable existential self-determination, which we undertake consciously or unconsciously, leads to opinions regarding the meaning of life. These opinions can be examined to see whether they are coherent, i.e. whether they are generally valid when examined more closely from the point of view of the agent. And it is precisely this cogency that is lacking in the view that the natural sciences have abolished the meaning of life because they have shown us that we have no meaning for the cosmos. On closer examination, this belief is unsatisfactory from the point of view of the agent. Nagel provides the following interesting justification for this.

It seems obvious to look for the meaning of life in something bigger than ourselves. We can look for it in God, the progress of humanity, the well-being of future generations, scientific progress, the success of the company we work for, or in sports. All of these meaningful projects transcend our individual lives in one way or another and seem to lend meaning to them precisely because of this transcendence. Let us call this **existential transcendence**: our own self-determination receives its meaning from something that transcends it and in which it is embedded. The existential philosopher Karl Jaspers speaks in this context of "the encompassing."[69] Thomas Nagel expresses it this way: "But a role in some larger enterprise cannot confer significance unless that enterprise is itself significant. And its significance must come back to what we can understand, or it will not even appear to give us what we are seeking."[70] The encompassing must therefore be something that itself has meaning, because otherwise it would no longer make sense from the agent's perspective to orient oneself to it. Consequently, it is no wonder that life appears to us as meaningless when we compare it with the vastness of the cosmos, which indeed has no existential meaning. The astronomical number of galaxies or the mathematical structure of space-time are worthy objects of reflection and teach us thereby that a scientist's life can be existentially meaningful. But the objects themselves do not confer existential meaning, which can be shown by a classical existentialist finger exercise. Let's imagine a more or less steep slope that we encounter during an alpine hike. If we are on a guided hike with Reinhold Messner (a famous alpinist), this slope will appear to us as a challenge to be overcome. In this context, the purpose of the hike is to evaluate the landscape in terms of the challenge it poses to our alpine skills. It is different when we are on a slow and relaxing walk with our grandparents. Then the slope might appear as an insurmountable obstacle to be avoided. The slope in itself, as Sartre would say, has no existential meaning; it has meaning only for us. Depending on how we determine ourselves and put our abilities to use, the meaning of the slope will vary.

It is the same with the cosmos and the universe. They can appear to us on the one hand as an ice-cold, dangerous and infinitely big desert of insignificance or, conversely, as a manifestation of a trace of the divine. In itself the cosmos and the universe have no existential meaning, but from this it does not follow that there is no existential meaning!

Nevertheless, the cosmos has a special meaning in the project of existential self-determination. For we locate ourselves in one way or another in a more or less comprehensive whole, by considering ourselves as part of nature for instance. As long as we can place ourselves in the external, transcendental standpoint from which we view our life in this same whole, what may seem to us at the moment to be of infinite importance appears to be less significant. Consider, for example, what people are experiencing at this very moment as you read these lines – from the greatest joys of embracing a loved one to a serious car accident, not to mention the terrible fact that, at the time of writing, many people in Germany alone are lying lonely, frightened and seriously ill with Covid-19 in an intensive care unit, as well as tens of thousands of victims of horrific wars in Ukraine and other countries in which we happen to be less interested (Syria, Yemen, etc.).

Whatever seems important, significant, and meaningful at a given moment may look comical or absurd when viewed from a cosmic angle, as Nagel points out:

> Reference to our small size and short lifespan and to the fact that all of mankind will eventually vanish without a trace are metaphors for the backward step which permits us to regard ourselves from without and to find the particular form of our lives curious and slightly surprising. By feigning a nebula's-eye view, we illustrate the capacity to see ourselves without presuppositions, as arbitrary, idiosyncratic, highly specific occupants of the world, one of countless possible forms of life.[71]

However, Nagel makes a mistake at this and other points in his important philosophical work. For, from the fact that we can look at ourselves from the point of view of a whole which embraces our life, it does not always follow that this appears to be comic, absurd, contingent and insignificant. Rather, we can also understand the cosmos as a manifestation of a trace of the divine or some other meaning-giving process. Insight into the vastness of the cosmos does not automatically trigger a feeling of meaninglessness and lack of significance. It can also be experienced as a revelation of the power of the human mind to comprehend the cosmos or as an expression of a divine power to create something that transcends our wildest imagination.

We can feel not only liberated, comical and absurd when considering the whole of the universe but also elevated and loved. Like other existentialists before him, Nagel can be accused of forgetting love at this point. In life's borderline situations, such as birth, falling in love, death and serious illness, we can experience the encompassing in its power as a foundation of meaning. It then seems to us as if everything that happens to us, perhaps even everything that happens to people at all, is wrapped in a mysterious meaning, as if even our everyday, banal actions had a deeper meaning. This existential experience, which everyone has gone through in some way, does not leave us as insignificant, meaningless creatures, but it sometimes shows us that our lives are not meaningless. The outside perspective on the cosmos is compatible with the inside view of the significance of what we do and what happens to us. We just have to realize that the outside perspective on the cosmos has to include our inside view, otherwise we miss something essential, namely ourselves.

This is not to glorify love or even pain. The same experiences that put our lives in the light of an encompassing meaning can also have the opposite effect. Death and serious illness should not be taken as a source of value, as they usually undermine the experience of meaning for those affected and, in particularly tragic situations, can mean that people never return to finding their lives meaningful again. Grief can be so severe that it crushes us – an existential fact that is given too little attention in the discussion about the Covid pandemic because we have become accustomed to taking note of the daily infection and death figures just like the weather report.

What this reflection shows is that the standpoint of the encompassing, which we can take up existentially in order to evaluate our life from there, does not automatically lead us to consider our lives as comical and absurd as Nagel thinks. It is on the contrary because of our capacity for existential transcendence that we are able to pose the question of the meaning of life.

From Mind Back to Nature

Ethics cannot be reduced to a procedure for deriving a socially acceptable and politically justifiable calculation of happiness for human animals. It is not only about putting pain and pleasure into an appropriate relation, i.e. to avoid pain and to strive for pleasure. There is no ethical calculus. The standpoint from which we can make value judgments at all presupposes an exercise of our capacities to live a life in light of a conception of who we are and who we want to be.

I bring these abilities together under the term **mind**.[72] The mental abilities of the human being depend on being introduced to the realm of agency under conditions of social freedom. Our life therefore has meaning only in connection with the lives of other people. Our existential self-determination is the basis of all concrete value judgments. In order to be able to make well-founded value judgments, we need social structures whose function is to train people in ethical judgment. This is exactly why the New Enlightenment calls for ethics for all (i.e., specifically, ethics classes beginning at primary school at the latest), so that we can exchange ideas together about the development of our social freedom and the meaning of life.[73]

Because, as human animals, we are incessantly confronted with our internal nature and its position in external nature (the environment), the life of mind is not an unclouded self-manifestation, decoupled from nature, but is involved in processes we can always only partially grasp. Our vulnerability, our being-an-animal, accordingly also shows itself when we reach the highest levels of human self-knowledge.

In the life of mind, mind and nature meet and become entangled in our lives. The entanglement of mind and nature is only partially explorable by natural science, for existential self-determination – i.e. mind – transcends nature. We merge completely with nature only when our mental life ends. As long as we live, we remain at a distance from internal and external nature, which does not mean that there are two independent substances or realities – mind and nature.[74] In being human, mind and nature actually hang together.

As living beings, our embodiment relies on processes of negentropy, i.e. on the fact that our organic, cell-based constitution endures despite environmental pressure. The environment (or nature) is always potentially threatening to our organism. The idea that we should somehow return to nature in order to overcome the existential and ecological crisis of humanity, as some postulate, overlooks the fact that the environment and nature are and always have been dangerous for us. Of course, it is only through and thanks to natural structures that we live. We feed on cell-based food and we breathe the air whose oxygen was produced over billions of years by bacteria and other life forms, long before we existed. And all the materials we use to build our homes are found on planet Earth as raw materials that were at some point part of life-cycles (again, bacteria, plants, lichens, etc., play an essential role).

It does not follow that we should subjugate, control, or dominate nature. For we cannot succeed in this. Sooner or later, every living thing succumbs to environmental pressure. Nature demands its right, and we dissolve. When and how this happens, we do not know. The pre-Socratic natural philosopher Anaximander expresses this in an effective dictum which has achieved fame as **the saying of Anaximander**: "And the things out of which birth comes about for beings, into these too their destruction happens, according to obligation: for they pay the penalty (*dike*) and retribution (*tisis*) to each other for their injustice (*adikia*) according to the order of time[.]"[75] As far as we know, our mind is bound to embodiment and has necessary physical, biochemical and neural (in a single word: natural) presuppositions. It is not, however, identical with these presuppositions. That is to say, nature is a necessary but not a sufficient condition for the existence of mind, otherwise everything that is natural would also be mental – and that is not the case.[76] Anaximander was one of the first to recognize that the stability of nature consists in its change, from which he drew the remarkable conclusion that the origin of all that is natural or all that exists cannot itself be identified with anything that is known to us. This origin he calls the unlimited (*to apeiron*).

In one particular respect, which will concern us more fully in the next part of the book, we are no more advanced today than Anaximander. Of course, we know much more about the concrete structures of natural change because, in modernity, our mathematical methods and our scientific instruments have advanced rapidly and in unprecedented ways. Viruses, cells, electrons, carbon and galaxies were as foreign to Anaximander as the exponential function or the nabla operator in vector analysis. But no matter how exactly we dissect nature and divide it meaningfully into recurring units, we do not have a comprehensive understanding or even an explanatory model of all natural processes – which is again a realization that we have attained thanks to modern mathematical, scientific and technological progress. It is therefore the task of the theory of science and epistemology to clarify this state of affairs.[77]

It is certain that humanity (as of 2022) is indeterminately far away from having deciphered nature in its entirety as explainable. We do not even know how far we have actually progressed relative to a final comprehensive theory of nature since Anaximander. It is accordingly imperative that we exercise epistemic modesty, especially in the field of natural science, despite all our impressive and gratifying successes.

---

*This should also include avoiding a widespread fallacy which I would like to call the **identity trap**. This fallacy deduces from the fact that mind occurs (so far as we know) only when certain natural, organic conditions are fulfilled, that these conditions are not only necessary but also sufficient presuppositions for mind. Then the mind would be identical with exactly these conditions.*

Strictly speaking, we do not even know whether the natural conditions for the mind are necessary. If there should be God, gods, angels, immortal souls or any of the other innumerable metaphysical entities in which the majority of humanity believes, one could not dismiss the idea that the meaning of life stems from contact with transcendence. To say it in the suitable poetic language of Friedrich Schiller:

> Friendless was the Mighty Lord of Earth,
> Felt a Want – so gave the Spirit birth,
> Mirror blest where His own glories shine! –
> Ne'er his Like has found that Being high, –
> Nought e'er gushes – save Infinity –
> From the Spirit-Region's Cup Divine![78]

It would be presumptuous to claim that we can exclude the existence of a transcendent sphere of reality because we know nature today more exactly than Schiller, Jesus Christ, the prophet Mohammed, the Buddha or some other religious genius. For scientific research does not deconstruct the mind but analyzes nature by means of the suitable methods that have developed in the different disciplines of the natural sciences.

Some of the results of so-called basic research reach the public because scientific research achieves results that can be technologically and politically exploited and are of drastic and groundbreaking importance for our community. There is nothing to be said against this in principle, as long as techno-scientific progress does not devastate the planet as an unleashed innovation, which still overwhelmingly takes place today. Today's over-engineered consumer society is heated up by fossil fuels, nuclear power and other energy sources, which have brought us to the edge of our self-destruction and have claimed many human lives. This is also due to the fact that basic scientific research as such does not contain any criteria as to whether and how its results should be assessed and technologically and politically utilized. This question cannot be answered by natural science but, if at all, only by the humanities and social sciences.

The New Enlightenment places this really very simple insight at the center of our task of shaping society, a task given each of us simply because we are capable of moral insight in different degrees. Each of us is familiar with different moral facts, because we all lead our own lives with specific challenges that only affect us and the people with whom we are connected by our actions.

As minded living beings we lead a *double life* without being amphibians (see p. 52). On the one hand, we have to survive, which presupposes the occurrence of organic processes without which we could not be embodied. Our bodies unquestionably depend on necessary natural conditions whose nature can be scientifically investigated. How vaccines support the immune system against pathogens is not a question for the humanities, or about the mind or the soul (although mental states are psychosomatically related to the immune system in complex ways). But we must not reduce ourselves to our survival, for the life of the mind demands more from us than making sure that our organism survives. If we were identical with our body, there would be no reason to care about others except to instrumentalize them for our own interests. Ethics would then at best be reduced to a tactic in the struggle for survival among more or less domesticated mammals. If this were the case, one could no longer really speak of ethics as an independent realm of thought, discourse and action.

Of course, this doesn't mean that the other life forms, some of which we also classify as mammals, are different from us in that they are identical with their bodies. Dolphins, gorillas, dogs and cats do many things that show they have feelings and states of consciousness similar to ours. I would not doubt that. We do not know how far consciousness and mind reach in nature; but we know that they are not limited to humans.

Mind and nature are intertwined in the meaning of life. As I have written elsewhere, "[t]he aim of human life is a good life. A good life is one in which we make ourselves responsible actors in the kingdom of ends and understand ourselves as living beings that are capable of higher, universal morality."[79] The good life does not take place only after earthly death (whatever, if anything, may take place then); the meaning of life must be tangible already here, in this life. In this respect, thinkers such as Nietzsche and Hägglund are right (see pp. 119ff.): the meaning of life cannot be a transcendent affair.

As minded living beings we are embodied. We are vulnerable animals who are aware of this fact. We cannot strip off and overcome our animality, our human animal-being, and we do not have to do so in order to find the meaning of life. It is instead necessary to recognize the natural conditions of our form of survival as the basis of our search for meaning, so that nature and mind intertwine in the meaning of life.

Thereby we are constitutively socialized, for mind develops in the form of human societies. In this way, we constantly change the natural foundations of our survival without illuminating the scope of this activity. Instead, in modernity (i.e. since an unleashed industrialization has cancelled the original Enlightenment pact of scientific-technological and human moral progress) we are caught in the hamster wheel of an acceleration that revolves around itself. The rotational speed of modernity tends to make us blind to what we do not know, so that we give in to our respective momentary impulses – a modern pathology that has been spreading inexorably, especially since the invention of the smartphone and the associated spread of social networks, news tickers, and so on.

The defense of the mind against the mindless impulses of a digitalized subjectivity, which gratefully surrenders to every momentary impression because it shows it the route to what it supposedly desires, is not an "unrealistic" statement from the ivory tower of a dusty philosophy. On the contrary, the reflection on ourselves as spiritual beings is the way to a different kind of **technology**, a different attitude towards technology, science and economy, which is based on the fact that we take seriously both the limits of knowledge and the limits of growth. Not everything is feasible; the human being remains mortal and vulnerable. We are and remain dependent on processes that we will never fully understand and will certainly not master and control.

In order to come to reflective awareness in an age of unprecedented acceleration, humanity needs an ethics of not-knowing that acknowledges what we know and do not know on even scientific grounds: we do not know enough about nature today to meet the human challenges of the climate crisis solely through our existing technological resources. From this I deduce that we need a fundamental reconsideration, a change of consciousness, which should also consist of cultivating sustainable cultural practices. As a simple example of this, one might refer quite naturally to the reading of books. Whoever reads a book doesn't destroy the planet (at least at the present time) and comes to reflective awareness, which is lost in the second-by-second cycle of the digital *attention economy*, as the urban planner Georg Franck has called it.[80] Contemporary "spiritual capitalism" that taps our attention and lures us in front of the screens again and again is not an energy-neutral, sustainable path to a brave new world. On the contrary, it's precisely part of that modern self-destructive form of life that we must overcome with a picture of the human being and nature that places human and therewith moral progress at the center of our self-determination.

PART III

TOWARDS AN ETHICS OF NOT-KNOWING

But nature is a stranger yet;
The ones that cite her most
Have never passed her haunted house,
Nor simplified her ghost.
To pity those that know her not
Is helped by the regret
That those who know her, know her less
The nearer her they get.

Emily Dickinson

As animals, we are part of nature. At the same time, as specifically minded living beings, we do not fit completely into nature. This dichotomy is expressed in our concept 'animal' and thus in our attitude towards life. Against this background, modern life appears as meaningless or as something to which we have to give meaning (which it does not have in and by itself). For how should there be meaning in nature, since it consists only of material-energetic processes, processes from which life merely arose at some point? It looks as if meaning can only be something that animals, and maybe only humans, produce in order to cope with the stress of their life and with the dangers of survival.

This modern attitude is often associated with Max Weber's observation of a "disenchantment of the world." Indeed, one reads in Weber:

> Thus the growing process of intellectualization and rationalization does *not* imply a growing understanding of the conditions under which we live. It means something quite different. It is the knowledge or the conviction that if *only we wished* to understand them we *could* do so at any time. It means that in principle, then, we are not ruled by mysterious, unpredictable forces, but that, on the contrary, we can in principle *control everything by means of calculation*. That in turn means the disenchantment of the world.[1]

According to Weber's analysis, the disenchantment of the world is a belief in the complete knowability and calculability of reality. On closer examination, this belief is a delusion. This is what the concluding part of this book is about.

The mind cannot be naturalized, i.e. completely explained by natural science and thus deciphered in the language of mathematical formulas. This is not to raise an objection against the attempt to collect as much scientific knowledge about the human being and its environment as possible and to use it for our welfare.

The highest stage of the mind lies in the fact that it grasps itself as such. The mind then attains the form of self-knowledge. To this self-knowledge belongs insight into the limitation of our knowledge of nature. Our knowledge of nature is thereby itself a manifestation of human mindedness: it is expressed by us as minded living beings, who thus become aware of their participation in a nature we can never fully control and explain. Nature, after all, is largely independent of our minds, in the sense that it is the way it is whether we like it or not. Nature corrects our assumptions about it, and we cannot make it up as we go along.

Part III is about not-knowing. Not-knowing belongs to knowing. For we know that there are lots of things we don't know, which is also a result of modern advances in knowledge in logic, mathematics and the natural sciences. In the humanities and social sciences this is acknowledged anyway, because they deal with processes of self-understanding and historical positioning of human beings which cannot be exhaustively grasped.[2]

Knowing is valuable. It consists in the knowledge of how reality is. Linguistically gifted animals such as we are can express their knowledge and communicate it to others. We make claims to knowledge and to truth. Whoever makes a knowledge claim can succeed or fail with it. Knowledge claims are subject to conditions of success.

In a modern knowledge-based society, we expect orientation from the sciences. We expect them to help us solve urgent social problems. This expectation recognizes the value of knowing. On the other hand, not-knowing has a bad reputation. It is equated with ignorance. We don't expect it to solve problems and provide answers.

But what I have described in these introductory remarks is not as self-evident as it may sound on closer examination. The problem-solving competence of the sciences does not stem from the fact that they work on solving urgent social problems. The sciences deal with different areas of reality, with different fields of sense. In the course of centuries, even millennia, they have developed a variety of methods to discover the truth in the fields of sense on which they focus. Scientific research is primarily guided by an interest in knowledge and truth, which consists in penetrating deeper and deeper into the specific matter of a science.

The deeper one penetrates into the matter of a science, the more precisely one grasps how much we as individuals and as "thinking collectives," as the microbiologist and epistemologist Ludwik Fleck has put it, still do not know.[3] The sciences are constantly pushing the boundaries of knowledge without ever being able to know how close they come to a comprehensive understanding of reality. For we do know that there is much we do not know. And we can also make assumptions and hypotheses about which facts we have not grasped and for what reasons. But we can't fathom the zone of not-knowing completely. Our human cognitive efforts, and so too the sciences, accordingly generate an "echo of not-knowing," as my Bonn colleague Wolfram Hogrebe has called it in a book dedicated to this issue.[4]

Against this background, it is naïve to think that we could or should align our human, historical, socio-political course with science. Insofar as an eminent singular such as "science" is admissible with respect to scientific theory, it is a collection of different methods that allow people to explore different areas of reality, whereby it turns out that there is much we do not know, even in the case of successful knowledge. Along with the stock of the objective knowledge of humanity – to which scientific knowledge from all disciplines contributes – our knowing of not-knowing also grows.

The sciences are not only guided by a pure interest in knowledge, however, although they are certainly also guided by this interest. Groundbreaking world-historical breakthroughs such as the development of quantum theory presuppose that humanity has developed basic mathematical structures and methods over hundreds of years. Their discovery presupposes a pure cognitive interest in logical-mathematical structures, without which there would be no modern natural science and so no modern technology either. But this pure interest in knowledge is embedded in social practices. For science is never done alone but always in exchange with others. Living in a society, being a social living being, always means for us human beings to be ready to be corrected. Scientists never exempt themselves from this.

The correctability of claims to knowledge and truth means that, while we can definitely know a lot of things, we have to be ready at the same time to acknowledge that we do not know what we do and don't know. The acquisition of knowledge is never finished; there can be no final and all-embracing knowledge of reality.[5] It doesn't follow from this that we can't know anything, only that we will never know everything. Thus we will never know what we know and what we don't know. Science remains piecemeal.

Because the sciences are embedded in social practices that intervene in nature and society, there are not only epistemological and scientific considerations of science but also an ethics of science. The actions of scientists are subject to ethical principles, unfortunately all too often disregarded, and this partly because research and animal ethics are unfortunately not sufficiently taken into account in science training. Thus the ethical principles of our everyday interactions also apply in the laboratory. Racism, misogyny and coarse humiliation are just as morally reprehensible at research institutions as they are in the subway. Additionally, scientists are known to have worked on technologies that gave rise to modern weapons of mass destruction and digital systems (think of social networks) which are also used today as weapons against liberal democracy. How we research nature and what technological products emerge from this research has massive ethical consequences, as we know (keywords: combustion engines, nuclear power, animal experiments, research on bioweapons, genetics research, etc.). The exploration of nature is never just a value-neutral observation of natural phenomena but always has desirable and undesirable dimensions, dimensions in which human value judgments and ideas are expressed.

One reason for the far-reaching failure of the scientific-technological enterprise to take account of ethics, the humanities and the social sciences is precisely that we trust the sciences in modernity with a promise of salvation because they have indeed either completely freed us from many evils or at least alleviated them. Thanks to the sciences, modern knowledge-based societies are highly equipped technologically, and this enables medical and social comfort (especially for people living in rich industrial societies).

It is now precisely this need for technological development, however, that is a driver of the problematic modern situation. The decoupling of scientific and technological progress from human and moral values and the association of science with a quasi-religious promise of eternal economic growth and eternal health on planet Earth have led to countless pathologies. Just think of the fierce debates about nuclear power, the atomic bomb and our fossil fuel society, which have recently been reignited. Without scientific-technological progress we would not be in the predicament of rapid human-made climate change, but we would also not have the standard of living we have in rich industrial nations. The ethics of knowing therefore does not automatically result from the fact that we demand even more scientific research and trust science even more than we already do. In short, the sciences are not the bringers of salvation and should not take over the function of religions. They belong to a completely different category of human knowledge and human self-knowledge than the latter.

It is therefore a modern mistake to believe that science and religion are in tension or even mutually exclusive – as if religion were a superstition which had to be enlightened and overcome by science. This view is itself an erroneous belief, for it misses the real character of the sciences as well as that of the religions at their core. Whatever the natural sciences will find out about the physically explorable universe or about the functioning of our nervous system, it can never be deduced that there is no God and no immortal soul. And the impressive results of evolutionary biology do not prove that there is no cycle of rebirths, as Hinduism and Buddhism believe. Hinduism and Buddhism are not just fasting or mindfulness exercises but make metaphysical claims concerning our salvation. In this soteriological function (**soteriology** = doctrine of redemption) they cannot and should not be replaced by the sciences.

Many socially influential sciences, such as physics, economics and law, intervene in natural and social reality in and through their research. The systems whose functioning they study can only be accessed by changing them through intervention (by setting up experiments and conducting studies). This is known in recent philosophy of science under the key word of interventionism.[6] At least since the groundbreaking discoveries of quantum theory, it has been known that the experimental setup of an experiment (a measurement) influences the behavior of the objects (e.g. electrons) being studied. In the case of the social sciences, the philosopher of law Christoph Möllers has shown that these can only explore norms that are explicitly sanctionable – that is, that they can only recognize something that must be elicited, so to speak, from implicit social reality.[7] For example, anyone who explores the voting behavior of a population through surveys also influences it in and through this exploration, e.g. by causing the respondents to think about their voting behavior in new ways. And, in the case of the legal sciences, it is true in any case that they not only research our behavior but also change it by making norms explicit, by writing them down, and by making them socio-politically effective, through state sanctions if necessary.

If one intervenes in nature and society out of an interest in knowledge, however – e.g. through an animal experiment, the development of the internet, research into biological weapons, the constitutional examination of a law, or a political opinion poll – this has ethical implications. Just as there is an ethics of science that deals with the question of the conditions under which scientists ought to achieve new results, there is also an ethics of not-knowing. This part of the book deals with its justification.

The ethics of not-knowing begins with the recognition of the fact that, thanks to our advances in knowledge, we know that there is much we do not know. We do not have an exhaustive or even all-encompassing understanding of reality. But it does not follow from this that we should give the time of day to fake news, the bullshit of "alternative facts," or worse forms of superstition and general science skepticism. For these modern phenomena are, rather, the expression of misguided knowledge claims. Whoever claims, despite completely clear data, that the Covid vaccinations did not curb the incidence of infection, that they even harmed more than they protected, does not represent ignorance but, rather, claims to know something (and is undoubtedly mistaken). The ethics of not-knowing is, on the contrary, based on the fact that we know many things; it does not question what we know factually. For that would be nonsense.

Instead, the ethics of not-knowing consists in working out to what extent nature and mind are disclosed to us and to what extent they remain alien and hidden. The ethics of not-knowing accepts that nature 'in and outside of us' is and remains deeply alien to us, as there is much about which we can at best only speculate. The exact extent of our knowing and not-knowing cannot be determined because the demarcation that distinguishes the unknown from the known cannot be made a priori, i.e. purely conceptually. At some point, with a view to nature, we are forced to speculate.

The things we can at best speculate about have immense consequences, however, which is why the ethics of not-knowing does not preach a mere **epistemic** (related to our knowledge, from Greek *epistêmê*) abstinence but recommends a deep respect for the otherness of nature, on which we as living beings depend without ever being able to control and domesticate it.

Here's a concrete example. Those who live as vegans because they believe that it is morally reprehensible to inflict pain on other living beings and kill them for food would be in serious trouble if it turned out that plants have consciousness and can feel pain. If the industrially organized consumption of broccoli and cauliflower or the cutting down of a fir tree for Christmas caused the same kind of suffering and pain as factory farming (or even more), this would change our environmental and food ethics drastically. Vegans would then not be allowed to eat anything at all if their reason for being vegan was to avoid inflicting pain (which is certainly not the only good reason for a vegan diet).

Most of humanity (both past and present) is convinced in one way or another that there is something higher or divine that expresses itself in nature and mind and transcends whatever we can grasp and know. Such a belief is based on the correct sense that we do not have a complete scientific worldview that can substitute for broader human, spiritual and religious needs. To believe that the sciences will one day explain reality sufficiently well in its totality to overcome such human needs, so that we could build a technocratic paradise on Earth, is itself an expression of a distorted form of this very spiritual need.[8]

It is therefore inevitable that we learn to deal with the gray areas of knowing and not-knowing if we want to handle the manifold crises in which humanity finds itself, both scientifically and socially. Therefore, in the name of a New Enlightenment, I call not for less science but, rather, for more science with the aim of advancing the self-enlightenment of humanity.[9]

This self-enlightenment involves our acceptance, guided by an epistemologically sustainable ethics, of the strangeness and otherness of nature as a source of non-knowing instead of chasing the mirage of a complete knowledge and mastery of nature. Whoever wants to dominate nature always wants to dominate other humans as well as the non-human life on our planet.

In sum: because there is an ethics of knowing, there must also be an ethics of not-knowing. For not-knowing plays a big role in human life. It expresses itself today as uncertainty in the face of the complexity of our crisis situation, which we can no longer overcome in a one-sidedly scientific-technological way. This concluding part of the book develops in an introductory and hopefully generally comprehensible way some foundations for a future ethics of not-knowing.

Nature, Environment, Universe

Reality appears to us as animals in a certain way. Because of our physiological equipment and genetic make-up, as well as our environmental adaptation in the course of our life, each individual becomes aware of different things at different times of their existence. We humans experience only a part of the radiation surrounding and penetrating us as color, sound, and so on, and thanks to our measuring instruments (microscopes, particle accelerators, computers, etc.) we know that nature in itself provides considerably more and different information than what appears to us directly on the level of our consciousness. We always see things, processes and persons only on a certain level of reality, which has been selected behind our backs (as it were) by the almost astronomically long processes of natural selection in the course of evolution. Our present condition as organisms is thus an expression of a complex natural history. The human being thus always grasps itself as an animal on the level of its mental penetration of reality.

Especially in the last two hundred years, the natural sciences have led to the realization that our organism dissects nature on many different levels in such a way that it introduces differences which then become conceptual divisions. Thus our organism distinguishes between nutrients and harmful substances, as well as between different wavelengths of radiation. We can process these distinctions on the levels of cultural concepts. We never experience reality as an all-encompassing whole but find ourselves in every moment of our conscious life in a subsystem, which in turn we break down into parts automatically and without fully conscious access to those processes of selection.

Sitting in front of me is the cereal bowl that just held my breakfast. I divide it into the cereal remains, the puddle of milk, the spoon that protrudes from the bowl, and the shining surface of the inside of the bowl, in which the light of my ceiling lamp is just now reflected. If I were now to examine the puddle of milk with a microscope, once again a completely different field of sense, whose objects are determined by the framing conditions of the microscope, would be revealed to me. In doing so, the microscope determines what is recognizable for me through it, without therefore producing the objects and facts I recognize in its field of sense. Bacteria and fungi that can be made visible to our eye through a microscope are not generated by the microscope. Yet those representations of bacteria and fungi that we can recognize on the user interface generated by and for us are.

Microscopes do not see milk better than the naked eye. They only see things on a different level, which is scaled by their adjustments. Neither what the microscope detects nor what the naked eye sees is the one reality in itself. Rather, both are realizations of details in their specific fields of sense. These fields of sense are in each case a reality in itself. There just isn't one reality in itself which we have to distinguish from a reality for us. How reality appears to us is quite simply part of reality, full stop. We don't look at reality from the outside.

Our knowledge of nature is always bound in one way or another to the fact that we grasp reality on some level through some instrument. In Greek, an instrument is called an *organon*. **Organs** (such as our organs of perception, eyes, ears, skin, etc.) are instruments shaped by evolutionary processes which capture a partially independent reality on a level assigned to them. An **organism** is a complex system of organs such that the more or less delimited whole of the system has a partial influence on how its organs develop. A whole human organism usually feeds itself in a more or less conscious way, which in turn influences our organs. The way we move (by doing or not doing sports, say) also influences our organs, creating complex cycles. If you do a lot of sports, you strengthen your heart, which in turn affects your calorie intake and leads to the strengthening of the organism.

Organisms are mereological cycles: the whole determines the behavior of the parts, and the behavior of the parts in turn determines the whole. On the one hand, the whole is more than its parts in an organism. But, on the other hand, the parts determine the way in which the whole transcends them. In the nature of an organism there are therefore different perspectives (the perspective of the liver as opposed to the heart, that of the brain as opposed to the skin, etc.), which the British physiologist Denis Noble has described in a series of worthwhile books in terms of a theory of **biological relativity**. Which part has an influence on which whole and vice versa depends on the perspective associated with an organ in the organism itself, the environment of which constitutes an environment for it in exactly the same way as the reality around us is experienced by us as a non-self or an external world.[10]

The non-self that is experienced is always only partially independent of the instruments that capture it. For the instruments can only capture it because they interact causally with it, i.e. because they intervene in the processes they observe. For example, the curtain of my study, which I have just opened to let in more sunlight, is not part of my self. It is a non-self. When I see the curtain, subatomic processes take place, which consist (among other things) of particles jumping back and forth between the curtain and me. Without such an interaction between the physical presence of my body and the curtain, there would be no perception at all. Things and processes which we consciously perceive are thus always in an interaction with us and are (imperceptibly) changed by those interactions.

We don't just know this since the quantum theory experiments, which clearly prove that the measuring process of quantum objects has a surprising causal influence on the measuring result. Quantum physics experiments such as the famous double-slit experiment clearly prove this. Roughly speaking, this classical experiment of quantum mechanics consists of elementary particles being shot at a wall, in front of which in turn is a wall in which there are two slits. Depending on whether one measures how the particles behave at the slits or not, they will produce either a wave or a particle pattern on the wall. We therefore know – and from many other experiments which prove this beyond doubt – that our measuring instruments have an influence on how elementary particles behave. The same is already true on the physiological level of perception, so we do not necessarily have to descend to the smallest levels of the universe to familiarize ourselves with quantum theory.

In order to perceive things, processes and persons that are in our environment, we have to appear in this environment itself as causally connected with it. We do not observe reality from a distant peephole. Our conscious mind does not use an immaterial telescope to grasp its surroundings. We are a part of every reality that we can observe.

Of course, the fact that perceptual reality is thus partially causally dependent on our being organisms in it does not mean that perception constructs a reality for itself or that it somehow falsifies reality. We are part of the causal *production* of nature. Thereby we select, in a largely non-conscious way, parts of nature that we consciously experience. Selection is not construction, and selection is certainly not unreal. It is based on causal processes of interaction with our environment.

The reality which we perceive is physically composed. Physics has clearly proven that we can only perceive the universe because we interact with it as bodies. Light particles (photons) have to hit our nerve endings in order for us to see anything. Conversely, we also change our environment by radiating heat, influencing the atmosphere through our breathing, and so on. We are never just passive spectators of natural events but, rather, causal components of the universe.

From the correct idea of selection, the idea that we select parts of nature for observation by our perception does not follow a thesis of the limitation of our knowledge of nature or even of the inevitable blindness and subjectivity of our perception. That our organism selects what and how it observes on different levels is compatible with the fact that thereby it grasps reality as it is on the observed level.

The fact that my cereal bowl catches my eye, that it stands out among the things on my desk, certainly also has evolutionary reasons. The color of the bowl as well as its contents can be explained by the fact that this color allows me to recognize the bowl against its background and by the fact that the contents are nutritious and taste good to me. Fully to explain this, of course, we would have to involve cultural, social and individual factors, since human nutrition can only be fully understood as a caloric intake that is useful for survival if we have no choice but to look for calories. One of the many reasons why we must make every effort to avoid extreme poverty and hunger is that, under conditions of extreme poverty, people are limited in their ability to participate in the cultural life of the mind. And so, in extreme cases, they are damaged in a crucial dimension of human life.

What our organs and instruments capture in each case is in principle never reality as a whole. They also do not grasp nature or even our life as a whole but, rather, perceive limited and open wholes which are constantly shifting. The whole of my field of vision changes incessantly on many levels, as in natural reality I perceive that everything is change and nothing is eternally stable.[11]

We have reached a point where conceptual definitions are necessary to untangle an epistemological knot in which we might otherwise get caught up.

An **environment** is a field of sense within which a life form can reproduce itself over several generations. Our environment is tailored to us as a life form, which does not mean that it is an already made, warm nest that we find. We are constantly changing our environment through our own reproduction. We not only adapt to it but also shape it through metabolism, so that we in turn must adapt to the environment we have helped to shape. **Ecology** explores processes of systemic adaptation as well as system production by living things on a scale that considers life forms and our causal environment. This is a mode of thought of the environmental sciences.

As already stated (see p. 113), the concept of environment was introduced by Jakob Johann von Uexküll in his work *Umwelt und Innenwelt der Tiere*. In that book, the concept contrasts with that of an external world, which contains everything that is intrinsically different from an animal. The environment is accordingly a field of sense tailored to an animal, its stimuli and interests, i.e. that part of reality in which animals are embedded or to which they are fitted, as Uexküll says. The environment is consciously and unconsciously shaped by animals; it is not just a niche they find and settle into. Uexküll illustrates this in a famous passage with reference to the sea urchin:

> In the sea urchin, to emphasize the essential again, it is not the unified impulse but the unified plan which pulls the whole environment of the animal into its organization. It selects from the useful and hostile objects of the environment those effects which are suitable as stimuli for a sea urchin. These stimuli correspond to graded receptive organs and centers, which respond differently to different stimuli and thereby excite the muscles that perform the movements envisaged by the plan.
>
> Thus, the sea urchin, too, is not exposed to a hostile outside world in which it leads a brutal struggle for existence. However, it lives in an environment which certainly contains harmfulness besides usefulness, but which fits its abilities to the last as if there were only one world and one sea urchin.[12]

The environment is only a part of natural reality. This part is the radius selected by stimulus–response patterns associated with an organism. Unlike the sea urchin, we humans transcend our environment by being able to perceive it as one of several environments. Humans are able to recognize that humans have a different environment than sea urchins and gazelles. We can therefore also ask the question as to how natural reality relates to these many environments – a question that urchins, gazelles, bacteria and nerve cells are not interested in.

In contrast to the environment, the **universe** is that field of sense which we adequately grasp only with our mathematical-physical models. It is not straightforwardly empirically experienceable. The universe is presented in the form of coordinate systems which resolve the natural events as accurately as possible into points on which forces act. All this can be described in principle in the languages of mathematics, which is why, for example, geometric relationships can be expressed analytically in order to make predictions about what measuring instruments would have to observe by means of sometimes gigantic processes of calculation. Theoretical physics can therefore develop experimental scenarios by mathematically supported thought experiments and also by setting up and solving equations, thanks to which theoretical assumptions can be experimentally tested. In this way, modern physics has developed into an admirable achievement of the human mind, no longer comprehensible in its entirety by any single individual.

The methodical kit of physics, together with its cycle of theory, model and experiment, allows us to make unprecedented progress in our understanding of the universe, which we now comprehend much better on all scales known to us than humanity ever did before. However, this certainly does not mean that we understand nature in its cycles, the environments of other life forms, or the rules of sustainable agriculture better than cultures that do not try to dominate nature through their exploration of the universe.

The major breakthroughs in modern physics did not occur in the early modern scientific revolution, although Galileo Galilei and Isaac Newton (to name but two of its heroes) did initiate the mathematical revolution. The greatest leaps begin in the second half of the nineteenth century, paving the way for quantum theory and relativity theory, which of course have in turn evolved since then – a story told in many other places.[13]

The universe is a mathematically representable structure, which on account of the limitation of our instruments we can grasp only to a limited extent. From our point of view, then, the universe is always a limited whole, and its limits could be further shifted by new physical findings in the future. In any case the particularly successful natural science of physics teaches us that there are limits to physical knowledge stemming from the fact that nature, as the target system of physical research, is dynamic. There are physical limits to what we can measure with our instruments in the first place, both because the universe is expanding (which we know through our mathematical, cosmological models) and because, as per our current knowledge, no information can be transmitted faster than the speed of light. Whatever is too distant from us in space-time to transmit information must remain in the dark for us. The same is true for so-called dark matter and dark energy, which can only be measured indirectly, and whose existence is inferred from theoretical results as well as from certain cosmological observations. Thanks to physics, we know that we will probably never know everything about nature through physics. Physics is capable of discovering its own limits without thereby somehow transgressing them. Whatever is on the other side of the limits of scientific knowledge is not indirectly knowable. That is the very point of thinking of physics as an empirical science.

In this context, **nature** is the target system of natural scientific research, paradigmatically of physics, chemistry and biology. While we can represent the universe in the language of mathematical formulas and corresponding models, we can also know nature in many other ways, e.g. through perception and cultural traditions. Nature and the universe are not the same. The universe is the part of nature that is representable in the mathematical terms of modern physics which do not exhaust everything there is in nature.

We do not know today whether there will one day be a state of knowledge that will allow us to express all of the life sciences in physical terms. If this were the case, the natural sciences as a whole could be reduced to physics. But at present we cannot determine in a scientific way whether there is ultimately only one natural science (physics). The natural sciences might remain insurmountably diverse, and this might be due to the fact that nature is essentially heterogeneous, that ultimately it consists of a multiplicity of fields of sense that cannot be reduced to one another.[14]

On the basis of his discovery of the existence of multiple environments (one for each animal), Uexküll argued that the very idea of an all-encompassing nature (a super-environment, as it were) does not make sense:

> If someone asks me doubtingly: "How can it be possible that my small subject with its frail environment should dictate its law to Sirius, which has been shining and shining for eons in infinite space in incredible size and distance?" I will answer him: "Pull out your markers for places and directions from the seemingly infinite space, and the whole space will collapse like a house of cards. And if you remove your symbols from Sirius, his existence will suddenly be cut off. Of course, all this applies only to you and your world. Nature will know how to create worlds even without you. There is no infinite, eternal and absolute world that encompasses all subjects – but there is an incredibly powerful nature that creates subjects with worlds, spaces and times according to its own free law."[15]

Uexküll describes nature in a truly ecological way. It is that which creates environments by producing living beings to which objects appear in different ways and to which they are adapted as living beings. Thereby he understands the mathematical models of physics, i.e. the coordinate systems we use to express objective physical relations for theory and experiment, as projection structures of our human environment. According to this account, the fact that our environment is understandable in terms of certain mathematical coordinate systems might be a provincial feature of our life form. We don't have to follow him here, as he underestimates the degree to which we humans can in principle transcend our niche. He believes that even as scientists we are only like sea urchins, exploring a reality into which we are fitted. But if this were true, how does Uexküll know that there is a "powerful nature" that produces an irreducible multiplicity of environments? Either this statement is just another projection of our local environment onto natural reality (the existence of which we, as higher urchins, could not really assure ourselves), or else it captures nature in a way that is independent of our provincial, human niche construction. But if living beings never grasp natural reality as it is in itself but always only the inside of their environmental enclosure, the thesis itself cannot be formulated as a claim about nature. He thus embroils himself in a subtle version of the projection thesis discussed in the first part of this book: because Uexküll subsumes human thinkers and sea urchins under the problematic concept 'animal,' he does not recognize that we can think far beyond our environment, for which purpose logic, mathematics, and other parts of our pure rationality are available to us. This pure rationality of ours which allows us to transcend our ecological niche, however, is not a bringer of salvation. On the contrary, by itself it is not fully in line with the ethical imperatives of practical reason; it brings about at least as much disaster as salvation. For this reason, the idea that society must be a purely rational (techno-scientific, economic) construct has caused much harm in modernity, as those on which this idea has been imposed in recent centuries through various waves of colonizing their life-worlds have had to experience.

In-Itself and For-Itself ...

Nature and mind hang together, but they are categorially different. That they are categorially different means that we can separate them conceptually without being entitled to conclude that they exist independently of each other. A categorial difference need not amount to a strong metaphysical one.

One essential difference between nature and mind is that nature is not interested in being known by us. It simply is as it is, and because of that it is not easy to know how it is. We needed millennia of human development before humans came to the ingenious idea to apply geometrical relations (above all the Pythagorean theorem, probably the most ingenious mathematical discovery of all time) to celestial phenomena via the observation of astronomical regularities. There was no deeper reason for the fusion of geometry and celestial observation. This simply happened at some point in human history in some human minds, even if this discovery is not completely coincidental but can to a certain extent be explained in terms of the history of ideas.

But we must not conclude from the contingent history of human knowledge that nature reveals itself to us. It neither reveals itself nor "keep[s] its secrets,"[16] as the natural philosopher Heraclitus conjectured. Nature simply is as it is. It is *in itself*, as this can be expressed in traditional philosophical language.

---

*Let us call this aspect of nature **being-in-itself**. Being-in-itself consists in the fact that nature is as we find it, without being tailored to our finding it or being interested in being found in any other way. The fact that nature is grasped by us partially in the form of a natural-scientific theory, i.e. as a universe, is contingent. Our perception of nature belongs in part to the in-itself, because it cannot take place without the natural preconditions of our organism. Perception is thus by no means 'purely subjective.'*

In short: it could easily have been the case that we humans had never developed deeper scientific knowledge through the unquestionably ingenious discovery of the connection between natural phenomena and mathematical models. While the natural processes we can grasp mathematically, in the form of the universe, proceed in a lawlike way (and in this sense are endowed with some form of necessity), this does not apply to the cognition of these natural processes. Certainly, we have been able to calculate the past and future position of the moon with the methods of the classical mechanics for centuries. But we cannot calculate at all what natural scientists will be able to calculate in the year 2055.

In contrast to nature, mind is something that essentially reveals itself to mind. Hegel, possibly the greatest philosopher of mind to date, calls this property of mind *manifestation*.[17] In contrast to the being-in-itself of nature, this can be described as the being-for-itself (*pour soi*) of mind, to use a Hegel-inspired category by Jean-Paul Sartre.

---

*The **being-for-itself** of the mind consists in the fact that the mind only exists by relating to itself in one way or another (such as in the form of consciousness). The existence of the mind does not remain unnoticed, because it exists as noticing itself. The mind essentially reveals itself to the mind; it cannot remain hidden.*

In contemporary philosophy of mind this is usually illustrated by reference to elementary states of consciousness. Whoever feels a stabbing pain cannot overlook it in the same way that the ancient Egyptians, for example, overlooked the distance to the nearest solar system, Proxima Centauri. Elementary states of consciousness do not go unnoticed; they cannot be overlooked. Mind exists only in its self-relation. A distant star observed by nobody, on the other hand, exists just like that, without any essential relation to a mind (with the possible exception of God, but that is another topic …).

Notice that it does not follow that mind or consciousness cannot deceive itself about itself. Mind manifests itself without thereby being an open book to itself. This is because mind is inseparably interwoven with nature. In addition to nature and mind there is their entanglement – i.e. a nature of mind as well as a mind of nature.[18] For the human being also belongs to nature. And since we are specifically minded living beings (those who can become aware of mind and nature as categories of what there is), nature recognizes itself in our self-knowledge as specifically minded. Nature and mind therefore do not belong to two completely separate fields of sense; they overlap in our minds.

Because the manifestation of mind has a natural basis, mind always shows itself only in a kind of distorting mirror. Its self-perception remains – at least in this life – dependent on embodiment. Since our body on some levels consists of mindless, anonymous processes, mind cannot fully grasp itself without recognizing its natural basis. But this natural basis is part of the in-itself, and it accordingly does not manifest itself. We all experienced this in-itself of our physical body every day in the Covid pandemic with respect to our body, for example, when wondering if we had been infected during a risky encounter. The infection does not take place consciously; one does not notice how the virions, the virus particles, penetrate our cells. The being-for-itself of mind is always partially non-transparent, because nature itself, seen from the point of view of the mind, has an indeterminably large part in mind. The in-itself and the for-itself do not form two metaphysical areas entirely isolated from each other but are, rather, dynamically interwoven. This entanglement does not undo their categorial distinctness.

*Every philosophy book needs at least one conceptual monstrosity; so, here goes. Let us call the entanglement of nature and mind, which occurs in a partially self-transparent way in the human being, **being-in-and-for-itself**. Minded life is on the one hand not completely knowable because of its natural basis (being-in-itself); on the other hand it is manifest – it expresses itself (being-for-itself). The human being is a being-in-and-for-itself which is capable of expressing this being-in-and-for-itself for itself (we just did it!).*

Strictly speaking, the situation is even more complicated, because we cannot know in principle whether we have fully cognized nature. No matter how far we get with natural knowledge, we will not meet a stop sign of nature or an ultimate enemy, as in a video game, after which the credits will run congratulating us on our final triumph. This means that we need to remain open to the possibility that the in-itself might turn out to be a for-itself (such as a divine self-revelation, as believers across religions as well as many theologians and metaphysicians have believed and argued over millennia). Having said that, on account of the limitations of our natural scientific knowledge, nature currently should, rather, strike us as truly being in-itself. Yet, to the extent to which our best physical theories remain empirical, this categorial claim concerning the status of nature is itself fallible and defeasible, so we might be wrong even in this part of our modern scientific view of reality (though I personally do not doubt it).

Is Science Fiction?

The difference between the being-in-itself of nature and the being-for-itself of mind is a basis for an ethics of not-knowing, i.e. an attitude of epistemic modesty, an attitude that allows us to recognize our knowledge claims as fallible and corrigible. Such epistemic modesty belongs to the standard repertoire of the scientific theory of falsificationism, which is accepted by most natural scientists. **Falsificationism** goes back to Karl Popper, one of the most influential philosophers of science of the last century. It states that scientific theories can never directly prove their truth but only establish either their falsity or rely on their non-falsity as long as not disproven.

The point of this idea is often illustrated by the **raven paradox**.[19] If we have observed only black ravens for a long time, we can formulate the statement: "All ravens are black." By doing this, we have ruled out the possibility of any raven being any other color. In this way, we have developed a fairly simple theory. This theory is falsified, thrown overboard, by someone showing us a single raven that is not black. That all ravens are black is therefore not confirmed definitively by any series of observations, no matter how good they may be, but is always valid only provisionally, i.e. as long as no non-black raven has been found. This threatens to amount to a paradox as soon as we generalize it to all scientific claims, as this entails that, in principle, all scientific knowledge claims to which we commit at any time could turn out to be wrong (which would have to be shown piecemeal, by finding counter-examples to every single one of them).

Falsificationism, which is widely accepted as an attitude of epistemic certainty in the natural sciences despite its known weaknesses (among which the threat of paradox), thus asserts

that a hypothesis can only be tested for falsity by observation. At best, one can show that something is not wrong – the way to truth remains blocked. … "Trial and error" is the credo of natural science – or, in the words of the physicist and philosopher Gerhard Vollmer, "We err upwards." At any time we work with models for the description of the world, which at best have proven to be "not wrong so far." We can only be sure about what has clearly been proved to be wrong. A much stonier way, as compared to mathematical proof.[20]

However, this argumentation overlooks the fact that not all propositions of the natural sciences are falsifiable in this way. If you state that not all ravens are black, you have not falsified the proposition that a given raven is black. That is, in order to falsify a statement or hypothesis, other statements and hypotheses must be established. If a statement or hypothesis has been falsified, this does not automatically mean that everything on which it is based has also been falsified. We cannot reasonably subject all statements of the natural sciences to the falsification test, which is why they progress cumulatively, i.e. findings build on findings without the entire edifice ever collapsing in on itself. It is possible in principle that we will one day overthrow much of what we today consider to be knowledge of nature. But it is impossible that all or even the greater majority of the natural scientific statements recognized as true today will turn out to be wrong. For there is not only scientific falsehood and non-falsity but also scientific truth. Therefore, it is possible to produce medicine and flight technology on the basis of scientific knowledge or to predict exactly where the moon will be in thousands of years and to calculate backward where it was thousands of years ago.

Every more or less comprehensive scientific theory known today (including the powerful standard model of particle physics) is a mixture of truth, falsity and non-falsity. This already results from the fact that there are no scientific theories without models, because models are not copies of reality on the scale 1:1 but representations of a partial area of nature, which are reduced to the essentials for the sake of theory construction. In nature, there are not just isolated elementary particles floating around that physicists somehow capture and measure. The very concept of a particle is already embedded in complex mathematical reasoning which belongs to the model construction of particle physics. We do not measure particles in the way in which, say, we see a car. We can only 'see' them in the context of models which are, of course, tied to detectors whose results in turn need to be interpreted from within the model. In order to register the natural, systemic relationships that occur in nature, one must analyze them and create conditions that normally do not occur in nature. Remember when you learned that feathers and cannonballs fall at the same rate in a vacuum? At that time, you did not learn that feathers and cannonballs fall at the same rate and thus, 'in truth,' have the same weight, which would be nonsense. Rather it was explained to you in elementary physics lessons that the regularities of the universe, which can be expressed in mathematical equations, can only be determined by intervening in nature. Without causal intervention, no models can be created that allow us to decompose nature into different fields and forces.[21] Feathers and cannonballs fall at the same rate only under certain conditions, and one can learn something about the universe from these conditions.

When analyzing a given scientific theory, it is anything but easy to determine which hypotheses, calculations and models apply to nature (and how exactly) or deviate from it in a way that nevertheless allows us to grasp nature sufficiently well to achieve a certain predictive or explanatory goal. Against this background, it is common in contemporary philosophy of science to understand scientific theories in analogy to fictions, a strategy which has a long tradition going back at least to Immanuel Kant.[22] In this context a fiction refers not to a pure invention but to a mixture of truth and a representation of nature which in part deliberately deviates from it, produced by simplification, idealization, abstraction and modeling.

Scientific fictionalism states (in a moderate form) that every scientific theory is a mixture of more or less consciously accepted half-truths, unrecognized falsehoods and non-falsities, but also of truths and precise measurements. The point of this consideration is that we cannot cleanly separate these elements. Hence we know that every scientific theory is falsifiable because it is literally false in some respects. This fact about scientific theories does not falsify them as a whole but, rather, makes scientific knowledge possible.

In his work *The Philosophy of "As If"*, the philosopher Hans Vaihinger, following Kant's idealistic epistemology, argued for the **extreme fictionalist thesis** that basically all entities and relations that play a role in the natural sciences (numbers, forces, elementary particles, causality, space, time, etc.) are fictions that the human mind constructs in order to form a simplified, necessarily distorted picture of the infinitely complex experiential reality that results from the stimulation of our senses.[23]

However, the argumentation designed to lead to this result breaks down if one considers that the assumption that our scientific theorizing is a construction on the basis of a flood of sensory stimuli (on the basis of which we construct a worldview) *itself* presupposes that one has gained scientific knowledge. For how does Vaihinger know that he receives sensory stimuli at all, stimuli he then makes understandable to himself by modeling them in terms of his fictionalism, i.e. by his own theory of complexity reduction? According to his own assumptions, this would already be a fiction in the sense of a distortion of reality. The idea of ourselves as animals, flooded by stimuli, rests on knowledge claims which, according to Vaihinger, are a fiction and thus do not correspond to reality – an ultimately untenable idea which fails in light of its self-application. The view that we actually only have our sense data or experiences at our disposal, on the basis of which we can then construct a more or less stable scientific view of the world, has for decades rightly been considered outdated and unjustifiable in the theory of knowledge and science. Another one of its fatal flaws is that it is based on so-called **epistemological foundationalism**, the assumption that there is an immediately given layer of elementary, conscious experiences and sense data that cannot be doubted any further. This view has been considered invalidated at least since the influential book *Empiricism and the Philosophy of Mind* by the American philosopher Wilfrid Sellars.[24] Sellars has shown that our cognitive apparatus can never process isolated, atomic sense impressions without placing them in a context in which they can be distinguished from one another. Here, too, a mereological argumentation comes into play, which does not reduce the whole of cognition to simple, isolated parts (atoms) but shows that the parts we can distinguish are already connected.

For our context, this means that there is no fundamental layer of human cognition (immune to revision) from which the natural world would have to be accessible to us, because every such layer already has a share in that very nature which is always partially inaccessible to us in its being-in-itself.

Falsificationism and moderate fictionalism are partial truths. For each describes an aspect of the scientific formation of knowledge. The natural sciences idealize by neglecting one or the other quantity even on an elementary level, rounding off somewhere after the decimal point or treating something as a mathematical point while in reality it might or even must have extension. This is not sloppiness (of which some pure mathematicians accuse physicists) but pragmatism, because one has to know which quantities can be neglected for the explanation of a natural phenomenon, even though, strictly speaking, they are mathematically significant. Vaihinger is not completely wrong when he reminds us that differential calculus (without which there would be no modern natural science) is based on the well-known trick that one divides almost by o, but stops dividing infinitely shortly before o. These are ingenious mathematical maneuvers without which there would be no modern natural science, but whether these ideas really correspond to nature (in which there is certainly no o) is another matter.

In short, we do not know in total which elements of the complex structures of the modern natural sciences correspond to natural reality and which parts are products of the human mind (of which products again some are true and others false). Nobody has an overview of the natural sciences and their logical-mathematical methods and foundations in their entirety. This is not a shortcoming but a recognizable fact unearthed by philosophy of science, thanks to which we can successfully conduct natural science.

Limits of Scientific Knowledge

Despite the impressive intellectual and technical achievements of our modern knowledge-based society, we do not know everything we don't know. The fact contained in the sentences just expressed is something that we know. Consequently, we know that there are limits of scientific knowledge.

The limits of scientific knowledge are dynamic, just as little fixed as anything else that we find in nature. For we constantly learn through trial and error what we did not know before and improve our assumptions in a cumulative process, which no one has in view all at once. Our not-knowing with respect to our not-knowing is thus an insurmountable, albeit adjustable, limit of scientific knowledge.

The adjustable, dynamic borders of scientific knowledge exist because nature is determined by its being-in-itself. It does not tell us what it is like, so we can never be completely sure we have grasped it for ourselves as it truly is in itself.

Mind you, it does not in any way follow from this fact that scientific research is a groping in the dark. Quite the contrary! If a given case of not-knowing is replaced by knowledge and progress thus takes place, then we learn how nature is in itself in a particular part. We by no means always replace an old error by a new one. So we have to subtract our own contribution to the exploration of nature out of the equations, as it were, which is possible in many cases. The fact that we can only recognize and explore nature by changing it in the process does not mean that we cannot know it as it is in itself.

Therefore we have to set our basic terms once again in oscillation, turning the thought spiral a little further. A few pages ago, I distinguished between nature and the universe. Do not worry, I do not take that back now. Nature is and remains the target system of scientific research, while the universe is the mathematically circumscribed area of nature more or less known to us so far.[25] However – and this is now the new step we are taking – the universe is a part of nature, namely that part which we can measure mathematically. Nature and universe are therefore by no means two completely separate fields of sense but dynamically interwoven with each other. This means that scientific research is advanced by researchers (by means of experiment and brain-power) who are themselves part of nature. The natural sciences do not observe nature from the outside. There is no "cosmic exile"[26] from which we determine nature's extension, as the philosopher of science Willard Van Orman Quine put it on the last pages of his main work *Word and Object*. Only as a part of nature can we observe it. In this sense, mind belongs to nature, as conversely nature belongs to mind. They contain each other reciprocally, because the part–whole relation (the mereology) of mind and nature cannot be understood as a material composition. Mind is not a part of nature in the way that the moon is a part of our solar system. And nature for its part is not a part of mind as the thought $E = mc^2$ (Einstein's famous formula, which makes claims about the equivalence of mass and energy and stands in pop culture for the achievements of modern physics) is a part of mind. For what it's worth, nature is 'out there,' not 'in here.' This claim needs to be taken with a pinch of salt precisely because nature also occurs 'in here,' where our mental processes take place.

The fact that the universe is part of nature also means that we can only scientifically measure those areas of nature for which we have technologies and thus instruments at our disposal. The biologists and philosophers Scott F. Gilbert, Jan Sapp and Alfred I. Tauber therefore rightly state: "We perceive only that part of nature that our technologies permit and, so too, our theories about nature are highly constrained by what our technologies enable us to observe."[27]

Of course, all this applies not only to physics but also to the life sciences. Our nervous system, of which our brain is a part, can also only be observed by using techniques such as modern imaging procedures that allow us to observe a living brain without having to interfere with it on a massive scale. Brain scanners that use functional magnetic resonance imaging (fMRI) to model what goes on in a human brain measure neural processes from a great distance. No one today is able to study complex neural activity at the level of individual neurons in time and space at a scale that even approaches 1:1. Imaging techniques produce models of the brain that are not 1:1 copies.

Neuroscience continues to advance, leading to the development of new techniques. For example, there are signs that quantum sensors could be used in the medium term to record neural currents more precisely than is currently possible by observing their magnetic fields, which should lead to major advances in knowledge.[28] However, we are far from being able to measure the entire brain in vivo, or even to correlate the majority of our mental states with neural processes – not to mention the philosophically significant problem that it does not follow from a correlation of neuronal patterns and mental states that one side of the correlation causes the other, even if the two form the two sides of the same coin (and not just two different coins).

This is all to say that, on the basis of scientific reasons alone, we cannot answer the question of the relationship between mental states (such as consciousness) and neural processes, and this is not only because we do not have the necessary technology. For while mental states and neural processes are correlated in manifold ways, they are categorially different. That means, in the best-case scenario, we can only discover correlations, from which it nonetheless cannot be deduced that brain states cause mental states (because then they would belong to two different realities, among which one, the brain states, would produce the other, the mental states, which is nonsense).

Another quite annoying limit of scientific knowledge, which will probably never be eliminated, arises from the complexity of the interconnection of different fields of sense. We can imagine this very concretely with the help of actual life during the pandemic, by now (unfortunately) quite familiar to all of us. On the one hand, a pandemic is an event that is defined epidemiologically. Whether a pandemic exists depends on how a microbiological pathogen (bacteria and viruses) spreads through the global human population. At this level, those medical sciences that study bacteria, viruses and their spread are in demand. On the other hand, of course, a pandemic is only a problem if it causes diseases that we notice. These diseases result from the fact that our immune systems are as it were naïve with respect to a particular pathogen, i.e. they don't yet have defense mechanisms that provide widespread immunity. Virology, epidemiology, bacteriology, immunology, cell biology, genetics, and so on, are responsible for making progress in knowledge at this level of reality.

But a virus that's dangerous from our human point of view not only affects cells into which it penetrates but also infects organs in this way, affecting our body as a whole. Our body is not only an object in the field of medical science. As a human body, it is part of social, economic, political, spiritual, intimate, or (in sum) cultural fields of sense. The human body is experienced in a certain, always socially determined way, which is why philosophy distinguishes between the **physical body** [*Körper*] that can be medically researched and the **living body** [*Leib*] that can be experienced from within.

The physical body is defined from the third-person perspective; it is what the doctor examines. The living body, conversely, can be experienced from the perspective of the first person; it always feels in a particular way, which certainly also holds for other living beings. The physical body and the living body belong to different fields of sense but not to entirely distinct worlds. On the contrary, they hang together.

The way in which we react to a pandemic socially changes its course. Societal reactions to a pandemic and the medical basis of a pandemic are not the same, but neither can they be understood independently of each other. One extreme reaction to this not purely medical fact of a pandemic is the most extensive possible lockdown, i.e. the dissolution of social relations through contact restrictions and prohibitions. The psychosocial, ecological and political effects of this are not, however, epidemiologically, virologically and immunologically comprehensible. Thus the effects of lockdown measures on the course of a pandemic are also underreported, because they consist not only in the fact that contacts are in fact restricted but also in the fact that school closures and social isolation cause damage to children, that companies go bankrupt, and that fundamental rights are curtailed, in some cases massively, rights which are supposed to protect us from an encroaching state. In addition, the lockdown measures have led to social groups rebelling against the lockdown, or even against any pandemic management, and to meetings in large groups for mask-free demonstrations, 'walks' and other gatherings, where they then become infected undetected, despite ordered contact restrictions. In short, a pandemic produces social effects. The discovery of a new pathogenic virus affects institutions (such as the WHO), which in turn affect society, which in turn affects the virus and its evolution – a complex cycle that can't be captured by epidemiology alone.

Nor can this cycle of effects be meaningfully explored solely through the sciences. For our socially orchestrated reactions to a pandemic happen to be part of it too: how we assess our individual and collective risk, what political views we have, and so on, form part of how the pathogen spreads. Researching a pandemic only from a scientific point of view and basing political decisions only on the scientific facts is a serious mistake that may even exacerbate the situation in the long run. Neither virology nor the natural sciences as a whole know what effects the many border closures and lockdowns have on the mental health of humanity as a whole and on the geopolitical situation, on which in turn the world economy depends and which must function well enough to fund the scientific research and health systems without which we would be helplessly exposed to the virus.

If we look only at the unattractive effects of the virus on the body, only at the human-animal population data, without considering the effects on politics, society and the individual psyche, we will not do justice to the phenomenon. It is different with bird flu, because birds do not react to its spread with measures that are medically planned and politically implemented. But human beings are neither birds nor pigs; they are minded living beings who can turn their own health into an object of research in order thus to intervene in their social structure.

Otherness – Towards an Ecological Ethics

In general, ethics, as a sub-discipline of philosophy, deals with moral facts. Moral facts are true answers to the question of what we should do or refrain from doing – just insofar as we are human beings.[29] For example, here are some simple moral facts: you should not throw bombs at kindergartens and you should help a person who has collapsed and cannot get up without help. Knowledge of moral facts is primarily anthropogenic, which means humans are the best known source of ethical knowledge so far. That ethics is *anthropogenic* does not automatically mean that it is also *anthropocentric*. From the fact that human beings can do ethics, recognize and acknowledge moral facts that are hidden from other living beings, it does not follow that we are entitled to subjugate the entire planet (which we are not even capable of doing). It does follow, however, that other living beings do not practice ethics and thus recognize moral facts only unsystematically or, as one says, instinctively.

Whether we like it or not, the scope of our responsibility far exceeds our obligations to other humans. The brutal factory farming system and the destructive exploitation of the habitats of other living beings are clearly morally reprehensible and, in the form currently practiced, even evil. These are also moral facts. For we could do otherwise; we are by no means dependent for our sheer survival on stuffing geese or forcing laying hens to produce eggs for us under degrading conditions.

It is important to note that it is precisely the uncontrolled growth of scientific-technological research and its economically driven application, not constrained by any thought-through ethics, that lies at the basis of the turbo-charged, runaway, predatory capitalism that has been afflicting the planet for more than two hundred years.

At this point, we encounter an important tradition in modern ethics associated with seminal works by Hannah Arendt, Hans Jonas, Emmanuel Levinas, Maurice Merleau-Ponty and Jacques Derrida. These philosophers have pointed out that radical alterity, i.e. radical otherness, is a fundamental principle of ethics.

Radical alterity *consists in the fact that the other in us and the others in general (other human beings, nature, other living beings, God and the gods, if one believes* *in them, and so on) are others precisely in that they have dimensions which we do not know. If the others in general and the other as such were an open book for us, completely knowable, they would not be really other but predictable repetitions of the same.*

The idea of radical alterity as a ground of ethics goes further than the widespread view that moral actions are altruistic, i.e. that we act in favor of others without regard to our own interests. This is because radical alterity is based on a view of otherness that does not presuppose that we already know how someone or something differs from us.

Others always surprise us, in a positive or a negative way. We have all often been deceived or disappointed by others. When we are disappointed by other people, a veil falls from our eyes, an illusion dissolves. Then suddenly we recognize the otherness of others, which was different from what we expected. Otherness is of course a source not only of disappointment but also of admiration, love, reverence and hospitality. How exactly it unfolds and reveals itself transcends our expectations.

All ethics begins with respect for otherness, with a politics of difference that recognizes that those living beings to whom we have moral obligations may be different from what we expect them to be and therefore deserve our respect. Alterity and animality, being other and being an animal, are interrelated and form the starting point for a reorientation of our understanding of nature. We need this more urgently than ever in an age of impending human self-extermination. Our own animal existence is alien to us, which is why we project it onto other living beings. But this is precisely how we miss them in their otherness, which should prompt us to revise both our attitude towards them and, as a result, our living together.

It is in the nature of human socialization that we are never able to see through and know all the people with whom we interact in such detail that our expectations and assessments coincide seamlessly with those of others. Therefore, there are no societies without conflicts of interest and without more or less extensive dissent.[30] Of course this applies to ourselves as individuals as well, as we are not completely transparent to ourselves and are therefore our own other. For this reason, there is not only an ethics that regulates relations between several persons but also an ethics that concerns our dealings with ourselves. We owe something not only to others because we share humanity with them but to ourselves as well.

Because others are not transparent to us, and therefore we cannot fully control them, it is part of ethics to work on seeing through our social projections as such. A **social projection** is an idea of the otherness of some given (factually existing or even altogether imagined) group in which the initial group uses its self-image to measure the foreign group against it. In doing so, the initial group does not take the position of the others seriously. Instead, they reduce it to that which is familiar. We know about this phenomenon from everyday life.

For example, a person from Hamburg imagines that people from Munich are more conservative than people from Hamburg. If the Hamburg person meets a Munich person, such an idea of the other will determine the former's expectations. Depending on how pronounced the prejudice against Munich people is, it will show up one way or another in her behavior. The foreign group of Munich residents is measured against the group to which the person from Hamburg feels she belongs. In doing so, she has of course overlooked (just like my own example, which is deliberately structured in this way) that the fact that some people in Hamburg are less conservative than some people in Munich (a true proposition, easily provable) does not imply that people in Hamburg are all progressive. There are also some Munich residents who are less conservative than some in Hamburg. Additionally, there are any number of political and social differences within Hamburg itself – just think of the two banks of the Alster River or the difference between Harburg and Eimsbüttel. All of this could in turn be differentiated at will. And in this way we could break down the prejudices on which my example thrives – an example which, as I said, I deliberately designed in such a way that you can see how easily we are guided by social projections in our thinking. You can easily adjust this German example to your own cultural context.

Not all social projections are reprehensible and dangerous prejudices. Particularly powerful social projections, much discussed because of their dangers and inherent violence, include nationalism, racism, xenophobia, and the various derogatory attitudes towards the gender aspects of human self-determination. The spectrum of pathological social projections that lead to violence against others, however, is much larger. In the Covid pandemic, for example, we got to know the manifold variations of **hygienism**, i.e. the reduction of others to potential sources of contagion and their supposed guilt for instances of infection (just think of the series of scapegoats, starting with the Chinese, children, travelers, foreigners from supposed and real virus-variant areas, the vaccination hustlers, right up to the unvaccinated, etc.).

In general, social conditions generate **stereotypes**, simplified and ultimately flawed explanations for why others are different.[31] Stereotypes come into play if we want to make sense of others' behavior when it seems incomprehensible to us. They reduce radical alterity to a scheme of expectation. They are paradoxical forms of assimilation of others to our own image of them. We then blame the actually radical otherness of others on something that we consider understandable – by classifying and experiencing others as 'typically' male, female, Bavarian, youthful, neoliberal, vegan, or whatever. When concrete explanations fail us, we take refuge in the abstractions of stereotypes. If this practice becomes socially effective, social projections arise that can become dangerous for others (most of whom then form a minority) if such projections are incorporated into regulations and laws or discharged in violence. Many of the ludicrously detailed Covid ordinances, which changed on a weekly basis (and sometimes even faster), have often had unintended but nevertheless harmful consequences for groups, groups that were either ignored or structurally discriminated against (such as the many children, or the South Africans refused entry to the EU in the weeks following the discovery of the Omicron variant). This is true even if the regulations were issued with the morally correct intention of saving lives and relieving the burden on the healthcare system.

In Part I of the book, I showed that the concept 'animal' contains parts of a massive (and for the so-called animals extremely harmful) social projection. We unconsciously project our own prejudices about nature, and about the 'animal' in us, onto the other living beings. The otherness of animals is not respected adequately as long as we measure it by how close or far away we think 'the animals' are when compared with us. So we currently believe that we can fight a virus or even a pandemic, which blinds us to the fact that "viruses are the most abundant entities in the biosphere,"[32] as a standard work of virology soberly and correctly states. You can protect yourself more or less well against viruses, but you can't fight them. This is not even to mention bacteria, which are clearly living organisms (in the case of viruses, this is controversial because they have no metabolism and cannot independently reproduce). "The average human body contains approximately 10^{13} cells, but these are outnumbered 10-fold by bacteria and as much as 100-fold by virus particles."[33]

The biosphere – in which we humans are an infinitesimally small part – is and remains deeply alien to us. The area that we know belongs to the universe, the scientifically explorable and explainable part of nature. But nature as a whole is not explorable, explicable and recognizable in this way.

Just as we can and should examine social ideas about others for their projective character in order to identify and correct pathological excesses of our social imagination, we must also consider the projective character of scientific models. These never match their target system 100 percent but always distort and simplify nature in some way. This is not an obstacle to knowledge as long as we recognize the limitations of scientific research. In our models of nature we always express in a partly conscious and partly unconscious way our views about the "human place in the cosmos," as the philosopher Max Scheler put it.

To recognize this limitation means to understand nature as a source of radical alterity and not to judge it only by human standards. The radical alterity of nature does not consist in the fact that it is beyond a "limit that completely separates"[34] us from it. It is not in itself unknowable, but it is capable of frustrating even our most well-founded claims to knowledge. This means that we are always in some way in error concerning nature, never able to eliminate all our errors. In concrete terms, thanks to cosmological research, we now have very good reasons to believe that the universe is 95 percent dark matter and dark energy, so called because we cannot measure them directly.

Nature is not "out there." Nature is not a wilderness from which we have cut ourselves off thanks to culture, civilization and technology. Nor is it true that modern humanity has somehow alienated itself from nature and now urgently needs to find a way back to nature in order to overcome the ecological crisis. That is wishful thinking. There is no "wild nature" on our planet in the area we inhabit that has not been influenced by human beings. In this sense, we have already distanced ourselves from nature many millennia ago, or, to be more precise, we have been distanced from it ever since we have existed. Nevertheless, nature is much closer to us than we would like. Wherever we are, nature also shows up. That is why we all deceive ourselves in one way or another, both about ourselves and about the fields of sense in which our lives take place.

The ecological ethics of the New Enlightenment is based on the realization that we simply could never master and domesticate nature even if we continued to do everything in our power to bring life under control. In the final analysis, nature is uncontrollable. For it is always different in some way other than it appears to us at the moment.

Under-Complex, Complex, Hyper-Complex

At this point, the New Enlightenment breaks with the promise of salvation of the eighteenth-century Enlightenment thinkers, which in retrospect was false and under-complex. They thought that a universal mathematical language could be developed that would allow all natural phenomena to be predicted. A famous thought experiment by the influential mathematician, physicist and astronomer Pierre-Simon, Marquis de Laplace, who even served as Napoleon Bonaparte's minister of the interior for a short time (Napoleon already knew Laplace as an examiner for the military), stands for this promise of salvation. In a much-quoted passage from his *A Philosophical Essay on Probabilities*, Laplace writes:

> We ought then to regard the present state of the universe as the effect of its anterior state and as the cause of the one which is to follow. Given for one instant an intelligence which could comprehend all the forces by which nature is animated and the respective situation of the beings who compose it – an intelligence sufficiently vast to submit these data to analysis – it would embrace in the same formula the movements of the greatest bodies of the universe and those of the lightest atom; for it, nothing would be uncertain and the future, as the past, would be present to its eyes.[35]

We know from the further development of modern physics (especially from relativity theory and quantum theory) that the mind imagined by Laplace cannot exist in nature. There are physical limits to the strictly deterministic predictability of natural phenomena – both in the smallest area of the universe known to us and on the maximum scale of the universe we can recognize as an overall cosmological system. Laplace ultimately overlooked the fact that we can only explore nature from an internal perspective, which is why nature can never be *présent à ses yeux* to any mind, as he has it, i.e. it can never lie open before the mind's eye. For the eye of this mind would either have to belong to nature itself, in which case the limits of knowledge would emerge, or it would have to be outside nature. But if it was outside nature (and thus 'eye' at best in a metaphorical sense), how would the data of nature (the *données*) reach it?

On the other hand, it is partly true, as Laplace continues, that the "human mind offers, in the perfection which it has been able to give to astronomy, a feeble idea of this intelligence."[36]

> All these efforts in the search for truth tend to lead it back continually to the vast intelligence which we have just mentioned, but from which it will always remain infinitely removed. This tendency, peculiar to the human race, is that which renders it superior to animals; and their progress in this respect distinguishes nations and ages and constitutes their true glory.[37]

This dramatic formulation reveals the skewed nature of Laplace's famous thought experiment, which expresses an incoherent overestimation of the human mind. For how does Laplace know that we remain infinitely distant from the perfect understanding to which the "past and future states of the system of the world"[38] are revealed while at the same time claiming that we constantly approach it? How can he compare a mind infinitely more perfect than ours with his own mind in order to measure the distance between us and the perfect mind's omniscience?

Moreover, it is striking that Laplace ascribes a superiority over animals to our weak mind, which is infinitely far from its desired perfection. In addition, he measures nations and ages by the scientific progress they have made. But if the goal of progress is and remains infinitely distant, in what way should this make us superior to animals or other nations that have contributed less to mathematical astronomy, as he believes? For an approach to the infinite always remains, by its very nature, infinitely far from its goal. That means the small steps taken by these nations, supposedly so great, are infinitesimal and can no longer count as a reason for Laplace's epistemic arrogance. At this point, from today's post-colonial perspective, it is easy to recognize the terrible prejudices against the "nations" subjugated and mistreated by the French empire Laplace represents.

In any case, Laplace imagines here at most a gradual superiority, one evoked by metaphors. Yet, who determines the concrete applications of the categories of superior and inferior? Is it an essential category to achieve the specific formal understanding of modern science (of which the mathematician, physicist and astronomer Laplace certainly had a lot)? Wouldn't a swallow, whether "European" or "African," laugh in a friendly way (if it could) and say: "You may be good at math, but you can get lost on a beer mat without a GPS, and I can fly unerringly from continent to continent. Now that's what I call understanding!"

It transpires that Laplace develops a rather incoherent fantasy of humanity and uses it as a yardstick to measure the value of animals as well as the value of other nations and ages – a typical case of the projection thesis formulated in the first part of this book.

The scientific view of the world that Laplace develops is under-complex. This is because it overlooks the interconnectedness of the system of acquiring knowledge with its target system, which in this case is nature. The term "complex" stems from the Latin (*cum* and *plicare*) and literally means "woven together." Complexity scientist Dirk Brockmann defines a "complex system" as a system that

> consists of different elements [that are] connected with each other and thereby, like wickerwork, form a structure that cannot be recognized in the one-cell elements. Just as, for example, no sweater is yet visible in the crochet stitch. "Complex" refers to the internal structure of a system or phenomenon, so it is an objective criterion. Whereas "complicated" always refers to the perceptive faculty of the observer. "Complicated" is subjective. Phenomena can be extraordinarily complex but uncomplicated.[39]

According to Brockmann, complexity research investigates "emergent behavior," i.e. behavior "whose structure cannot be deduced from the study of the individual elements."[40] If emergent behavior in this sense exists in reality ("objectively"), Laplace's perfect mind fails to recognize it because it only knows the position of the individual elements. If nature comprises complex phenomena at all, this already shows that its temporal unfolding (its future states) cannot be precisely predicted, even if complexity research itself is making impressive progress, progress Brockmann reports in his worthwhile book.

However, it is not enough to expand our scientific models by acknowledging the complexity of nature – even if this is an important step in the right direction. Brockmann rightly points out that the real challenge with a complex system involving us is to take humanities and social science findings into account. There are connections between biological processes and social processes (because we are social animals), and the latter cannot be reduced to their biological dimension. Natural phenomena are complex not just for objective reasons. Subjectivity also belongs at least partially to nature, as we are embodied as minded living beings. Mental and social processes also belong to nature and are related to it in a complex way. Apart from under-complex individual things and complex systems, there is also a hyper-complex layer of reality in nature.

---

*The **hyper-complexity** of nature consists in the fact that mind and nature are interwoven (com-plex) but still categorially distinct from each other, and so not the same. There is thus no comprehensive scientific standpoint from which we can explore the web of mind and nature that we represent ourselves as human beings.*

Because we can know and explore both nature and mind from within, we must acknowledge that we will never be able to assemble them into a single coherent whole. We can explore the interconnectedness of mind and nature from a variety of scientific standpoints, but in doing so we cannot develop a comprehensive discipline that describes all natural–mental phenomena in a single theoretical language. For any attempt to develop a universal theoretical language describing the nature–mind relationship will of necessity use either nature or the mind as a paradigm from which to determine the other side of the equation. Mind cannot be reduced to nature and nature cannot be reduced to mind.[41]

Homo sapiens, *or, The Wise Words of Socrates*

The human being is capable of wisdom, and so Carl Linnaeus rightly called us *Homo sapiens.* This, our capacity of wisdom, is not exhausted by the fact that we can control and modify our conditions of survival by science and technology. Wisdom does not boil down to applied science and technology. Any view which reduces wisdom to science and technology results in an instrumental attitude towards nature, which was rightly identified as a source of human self-destruction by the critical theory of Theodor W. Adorno and Max Horkheimer, in their seminal work *Dialectic of Enlightenment.*[42] I know I have said this a few times already, but it is a core statement that bears repeating like a mantra: techno-scientific progress decoupled from moral insight leads to the devastation of the planet. Accordingly, in response to environmental devastation, instead of more science and technology, we first need a radical rethinking of our finitude and limitations, a new account of wisdom.

The concept of wisdom to which Carl Linnaeus referred with his definition of the human being as *Homo sapiens* goes back to the Delphic oracle. According to Plato, the oracle is supposed to have identified Socrates as the "wisest of all men."[43] Socrates reports on this in the so-called *Apology of Socrates*, i.e. in his defense speech before the Athenian court, which is about to sentence him to death for blasphemy and corrupting the youth by introducing new, false gods.

In this context Socrates utters the famous saying that has entered cultural memory as "I know that I know nothing." Precisely in this knowledge of his not-knowing lies his wisdom, an apparent paradox that can be resolved on closer examination. For Socrates does not contradict himself with his famous saying. What he says in Greek can be translated without contradiction as follows:

The wise words of Socrates: *"I have become aware that I know nothing"* (emautôi synêidê oudén epistamenôi).[44]

The verb *syneidenai*, with which Socrates expresses his self-awareness, means "to know with." This knowing-with differs from scientific knowledge, *epistêmê*. The verb *syneidenai* later became the noun *syneidêsis*, which was translated into Latin as *con-scientia* (conscience). Long story short, Socrates is the discoverer of conscience as a form of self-knowledge distinct from the knowledge of experts. He recognizes the wisdom of conscience in the fact that knowledge of the good (i.e., of what we ought to do for moral reasons) cannot be obtained from an observer's perspective. Or, as a popular German saying goes, there is no good unless you do it [es gibt nichts Gutes, außer man tut es].

The good is recognized by trying to do it, and so it is bound to the reality of our action. Ethics is therefore practical knowledge, which is not only about what we should do or refrain from doing, but where insight should motivate corresponding action. We should not do everything we can do.

Moral conscience as a guide to ethics is not an infallible source of knowledge of moral facts. In our lives, it seldom comes forward in the form of absolute certainty but, rather, worries and irritates us by showing us that we can also be wrong. Conscience makes us aware that others could be right, to vary a formulation of the philosopher Hans-Georg Gadamer.[45]

At the same time, conscience is certainly a source of knowledge because it disturbs us in our often too firmly anchored judgments and prejudices – just as Socrates irritated the processes of Athenian democracy by annoying the experts with fundamental questions concerning their knowledge (questions which they could not answer). Socrates thus repeatedly puts his finger on the human hubris of wanting to get a grip on nature and society through expert knowledge alone.

In the *Apology*, Socrates refers to the *daimonion*, an inner voice, which he classifies as something divine. This does not mean that he considers conscience as a voice of God, as it were. Rather, he understands it as something that gives us orientation, without being reducible to a mere opinion. Wisdom consists in beginning to follow one's own conscience without falling prey to the error that it is an infallible voice. It depends, rather, on keeping oneself open to the fact that others can also be right, in order to view reality from many perspectives. Only in this way can we grasp the complexity of nature and mind, which we can approach through practices of wisdom but can never bring totally under control.

Opinions, Knowledge and the Idea of the Good

As is well known, *philo-sophia* means love of wisdom. Plato distinguishes between philosophy and sophistry, which consists of making sure that a political speaker is right in the heat of the moment by using arguments that are sometimes clever and sometimes openly invalid. In this context, we owe Plato and Socrates (historically, it is not easy to distinguish which of them made which contribution to Plato's ideas) the insight that opinion (*doxa*) and knowledge (*epistêmê*) are conceptually distinct.

An opinion is an assumption that can be true or false. We have countless opinions about countless topics. As social beings, we exchange opinions and come together in groups through our opinions. An opinion is a kind of belief or faith. It can be more or less strong. We can easily dissociate ourselves from some opinions, while others are firmly rooted. Public debates are often about issues that affect our political community. In order to discuss what we want to do and what we should do in the face of a *res publica*, i.e. a matter that concerns everyone, it is important to raise, compare and weigh opinions.

Not all opinions are equally good, however. There are various criteria by which we can classify opinions. In public debates, opinions prevail based on their strength, which can be measured in sociological, economic, legal and political terms. Opinions form a field of forces that is by no means merely an irrational struggle in which the strongest prevail. In the democratic constitutional state, many procedures have been developed to make minority opinions heard, to deprive loudly expressed opinions of their dominance, and to examine them from many perspectives. These include a pluralistic media landscape and a well-positioned political opposition. It is essential for the space of public opinion that there are complex balancing mechanisms in the field of forces, mechanisms not manageable by any one person, such that there are no easy paths to attain and maintain a monopoly on opinion in the long run.

The space of opinion follows laws that are complex along a multitude of dimensions. Governments research the opinions of voters in order to draw conclusions about what the population wants and how the will of the majority compares to that of minorities. This opinion research in turn influences the opinions of the population, which actively participates in surveys or speaks out in other ways.

Social structures are measured against a set of normative systems. In general, **normativity** refers to the fact that our thinking and acting (thinking is strictly speaking also a form of acting) is classified as correct or incorrect. Correct action is action that corresponds to a norm that exists either implicitly as an expectation or explicitly as a formulated rule.[46] Rules of a game, for example, are norms. Anyone who plays *Settlers of Catan* knows at some point what goods can be exchanged and how to collect goods in order to build a new settlement. This is written in the game instructions, but you have to learn it again in the game itself. As Aristotle knew, "the things we have to learn before we can do them, we learn by doing them."[47] The rules themselves determine whether you make a right or a wrong move, whether you like it or not. So when you play *Settlers of Catan*, you are doing something that is normative. That is true for a lot of what we do, but of course not for everything. There are no norms for how to be bored or how best to stare out the window. Not all human activity is normative, then, but a lot of it is.

Normativity is a social phenomenon. One cannot imagine norms without others observing whether one acts according to them. Even in cases of action that is hidden from others, in private, intimate and secret contexts, one follows rules that exist only because social practices exist. Put a different way, people never do anything that other people do not do.[48]

Back to Socrates and Plato. At this point in the philosophical reflection, the original Platonic or Socratic question arises as to whether there is a normativity that divides all opinions and all human social practice into correct and incorrect actions – regardless of whether the actors like it or not. This would mean that, even if an entire society had a unanimous opinion regarding certain actions, there would still be an evaluative standard that classifies their actions and thinking as right or wrong regardless of this opinion. So, for example, if more than 80 percent of the Russian population actually think that Putin's war of aggression against Ukraine is morally right, then a large majority of society would be mistaken in this case. The standard of evaluation thus exceeds the evaluation guidelines that society has implicitly or explicitly given itself. Let us call this **transcendent normativity**.

It is precisely with such normativity that ethics is concerned, with a normativity that cannot be replaced by any social opinion poll, by any court of law, by any government, by any institution, by anything that human beings could do and think. Plato's name for this normativity is the **idea of the good**, and his whole work can be read as an argumentation for the fact that the idea of the good casts an evaluative light on human society whether we like it or not.[49] Thus Plato also compares the idea of the good to the sun, without which nothing would be visible and recognizable and without which, of course, there would be no life.

Normativity fans out into different fields of sense. Legal normativity functions differently from that of sports clubs or board games; logical-mathematical normativity differently from that of love relationships; party-political normativity differently from that of media coverage of political negotiation; aesthetic normativity in the evaluation of artworks differently from that of a kindergarten group; and so on. One way of representing society is to analyze it in terms of the interaction of different fields of normativity.

For as long as ethics has existed as a systematic discipline (starting with Socrates and Plato), it has been based on the recognition of a transcendent normativity, one that exceeds the events of society as a whole along with its complex normative fields of force.

In concrete terms, this means that there are unconditional norms or values that cannot be changed by any social consensus. Examples of this can easily be found: rape, child molestation, dumping toxic substances into the oceans, human trafficking, genocide and mass murder, sadistic cruelty towards humans or other living beings, anti-Semitism, and other forms of racism are all morally reprehensible or evil. A fair distribution of goods, loving care for people and other living beings in order to strengthen them in their self-development, or the saving of a human life (if one does not put oneself in danger) are all conversely morally required or good. Even if all people (including those involved) agreed that something we classify as radically evil is acceptable or even desirable, they would be wrong and would commit an objective error. Thus, quite generally, human trafficking is morally reprehensible. The system of human trafficking is immoral, although it doesn't follow that nobody or fewer people should come to Europe. It follows, rather, that it is our moral duty to find ways to accept people into Europe in a morally justifiable way and to form a value-guided foreign policy that leads as far as possible to cooperating on an equal footing, especially with our fellow human beings from Africa. Of course, this is happening to some extent and is part of any legitimate Africa policy. But it is obviously not enough, which is why there are daily tragedies in the Mediterranean and on other external borders of the EU, all of which we are reporting much less at the moment because we are struggling with our health problems in the pandemic, and because now the terrible war in Ukraine is also capturing our attention. Europe's Africa policies speak of a colossal moral failure and are in urgent need of revision. The conscience of every single person, if they were honest, would revolt against the fact that we in the rich West are constantly boosting ourselves against new virus variants and withholding vaccines and know-how from the African continent (keyword: patents). Not to mention the fact that we are worried about energy and raw material shortages, while a lack of imports because of the war in Ukraine is already causing famine in some African countries.

Socrates' conscience expresses the fact that one can open oneself to the light of transcendent normativity, to the idea of the good. Socrates and Plato therefore regard conscience as "divine" because it exceeds and corrects people's opinions. It is beyond all arbitrariness and, therefore, also beyond all arbitrariness on the part of the state – a fact that is reflected in Germany's Basic Law, with its commitment to human dignity. In view of the world-historical experiences of twentieth-century atrocities (in my country, dramatically associated with the National Socialist reign of terror and its crimes against humanity), the Basic Law represents an appropriate way of recognizing a normativity that acts as something that we should absolutely acknowledge.

Ethics is based not only on opinion or the formation of opinion but also on knowledge. In contrast to the formation of opinion, knowledge (*epistêmê*) consists of our attempting to grasp facts that are independent of our beliefs. The molecular biological structure of a virus, the acceleration of the Earth, the solution of a differential equation, and the number of beds in a hospital are all examples of facts that exist and that we cannot change through our opinions alone. Whether we like it or not, the coronavirus that has kept us on edge since the end of 2019 is just too dangerous to ignore and simply carry on living as before. And, whether we like it or not, the cross-product of two vectors is perpendicular to the plane which they span, and so on.

Knowing something does not necessarily mean having the feeling of absolute certainty. One can know many things without knowing that one knows them. The categorial difference between knowledge or science (as *epistêmê* can also be translated) and other forms of belief lies, rather, in the way we tie knowledge to reasons, to *logos*, as Plato calls it.

Knowledge, according to Plato, is a true opinion that is tied to a logos, i.e. a methodically demanding justification. Whoever makes a claim to knowledge thereby claims to be able to secure what they believe to know in the space of reasons, as one says in contemporary philosophy, and to defend it against objections.[50]

The exchange of reasons is not only a battle of opinions, as the goal of the exchange is the truth or the facts, and these are independent of this exchange. Thus while the sciences admittedly do not exist without social practices embedded in a variety of normativities, they are not identical with such practices. For the social practices serve to find the truth.

The sciences are epistemically modest insofar as they are based on a demanding concept of knowledge. After all, making a claim to knowledge means being open to being in error. The transcendent voice of knowledge is not some error-free revelation on which philosophers can call. Philosophy and therewith ethics (a sub-discipline of philosophy, just as mechanics and optics are sub-disciplines of physics) therefore differ from prophecy or error-free revelations.

We owe Socrates the insight that there is a conscience, known to us as an inner voice, which calls out to warn us that we might be on a wrong path. This inner voice rarely speaks to us so clearly that we would simply have to implement what it whispers to us. If this were the case, we would not be free in our own moral judgment. Formation of conscience is always a process that triggers an inner dialectic. Otherwise, conscience would be only a moral instinct. We must treat conscience with extreme caution and understand its call as an occasion to think very carefully about the situation in which we find ourselves. This presupposes being ready to listen to others. It is conscience that distinguishes us from other beings who do not participate in moral progress, but who can act well instinctively and, in this way, grasp moral facts.

Openness to others reaches its limits, however, where radical evil is done. That is why it is important for a functioning society to establish normative, ethically thought-through cornerstones and integrate them into moral progress. We must not regress from the insights that people are fundamentally equal in their dignity; that women must not be excluded from social self-determination in any relevant respect (as well as people with various forms of gender self-determination that do not fit into simple, binary representations of male/female roles); that racism is a reprehensible form of thought and action; that children must be granted rights that are respected by the state, even if children do not yet have the right to vote (which is incidentally a massive moral flaw in our democracy); and so on. In short, we need an ethics that both allows us to make claims to knowledge and recognizes that these claims to knowledge can fail, i.e. that we are fallible – we can fail – in respect to ethical questions. This means that we must also deal ethically with not-knowing, which is a decisive factor in complex situations of action in which many of the moral facts are not obvious.

Moral Reality and Ethical Facts

Because ethics makes knowledge claims, it must be open to counter-arguments. But this openness has limits. The lower limit of ethics, without which it cannot exist as a science, is the assumption that there are moral facts which transcend our opinions in the first place. That this is not a wild or even a speculative metaphysical opinion can hopefully be seen better if we realize what a fact is.

In general, a **fact** is a true answer to a meaningful question.[51] So the answer to the question "Was Paris the capital of France on April 8, 2022?" is "Yes." Hence, the circumstance that Paris was the capital of France on April 8, 2022, is a fact. Of course, facts also exist for which we have not yet been able to formulate a question. All human languages taken together do not exhaust the range of facts. The facts do not form a totality which we could represent linguistically or comprehensively in any other way. As Hamlet remarks to Horatio in the famous fifth scene in Act I: "There are more things in heaven and earth, / than are dreamt of in your philosophy." That there are indeterminately more facts than factually answerable questions is the thesis of **ontological fact realism**. This thesis assumes that the realm of facts goes indeterminately far beyond what we can express linguistically.

Now, a **moral fact** is a fact that is an answer to an ethical question. An ethical question concerns our actions. It asks what we should or should not do merely insofar as we are human beings. Moral facts have an imperative character; this means they prescribe something (an action or its omission). Thus moral facts exist only in factual contexts of action. What one should do or refrain from doing as a human being is a question of the context of action and therefore in many cases unfortunately not easily answerable.

In the history of ethics, there is a multitude of proposals on how to formulate a universal rule from which it would be possible to derive the moral facts in individual cases. These include such famous approaches as Kant's categorical imperative, the Golden Rule ("Do not do to others what you would not have them do to you") and the maxim of the utilitarians, who want to calculate how an action affects others in order to deduce which option entails the greatest good for the greatest number, or the least harm.

These rules are at best initial approaches to ethical orientation, however. They cover important partial aspects of ethical reflection and can serve as guard rails. But it does not follow from the categorical imperative, from the Golden Rule, or from utilitarian calculations how long a lockdown in a pandemic should be, how many people one is allowed to meet in view of a coronavirus wave on New Year's Eve, whether one should become vegan, whether the new construction of nuclear power plants is better than the deployment of wind turbines at mitigating the moral and general humane damage of human-made climate change, and so on.

Our everyday life is permeated by particular moral questions. How do you settle a dispute with a colleague? How do you deal with the serious illness of a loved one? When and how do we explain sexuality to our children? Which job offer should we accept, and how does this affect the other social relationships in which we are involved? In a dramatic way, the Covid pandemic has brought home to us how many of our everyday micro-decisions carry moral weight, from which it by no means follows that anyone in the current state of affairs could have simple yet comprehensive moral answers to what are sometimes extremely complex questions. The same applies to climate change and the ethically appropriate design of the inevitable socio-ecological transformation that will ideally lead from the pathologies of a consumer society to a more humane and thus ever more ethically thought-through form of community. Moral progress does not happen automatically in this process. This is illustrated by the severe regression of the aggressive war on Ukraine, against which large parts of humanity are in turn rebelling in order to support Ukraine as best they can in the face of a terrible tragedy.

> *The fact that ethics cannot be exhaustively presented or reduced to an all-encompassing rule is due to the fact that the reality of our actions is complex.*

At this point, we need to make a further distinction. In general, facts exist whether or not we can adequately name, grasp and recognize them. This is fact realism. In the ethical case, the peculiarity is that the moral facts cannot be radically different from what we categorize them to be. It makes no sense to think that it is a moral fact that we should kill all Covid-positive people in order to end the pandemic. Although this could be epidemiologically expedient (and is accepted for other living beings during epidemics, which is morally questionable by the way), it is clearly morally reprehensible, even evil. Fortunately, this is why no one considers it. The moral facts are knowable in principle; they are addressed to us. There can be no moral facts that we cannot know in principle, and this makes moral facts different from physical facts. Many questions of physics are forever and in principle unanswerable, because it may well be that there are universes that leave no causal traces in our universe and therefore remain unrecognizable to us. In these universes, largely or even fundamentally different laws of nature may apply.

But it is meaningless to assume that there can be moral universes in which completely different laws apply – universes in which, for example, it would be morally commanded to torture as many people as possible for no further reason. If moral facts are addressed to us as human beings, there can be no moral order which would be radically different from everything we consider morally meaningful so far. The idea of the good is transcendent insofar as it cannot be reduced to any given social formation.

The version of moral fact realism that I propose here is therefore paired with **epistemic moral realism**, i.e. with the thesis that we can recognize moral reality and to a certain extent have always recognized it correctly. The light of ethical reason, which allows us to recognize moral facts and thus make them socially and practically effective, has dawned at some point in human history. We do not know when, where and how, but the assumption that this began on the African continent is as plausible here as elsewhere in paleo-anthropology. Alongside ancient

China, Egypt (and thus the African cultural area) is ultimately the first well-documented advanced civilization that achieved excellence in all areas of human thought and action (unfortunately also at the level of slave-owning, exploitation and the autocratic subjugation of people, which are by no means "European" inventions). Of course, ethics does not require a sedentary advanced civilization or written records, but it is not easy to determine the moral level of groups of people for whom no records exist (there are of course still non-sedentary groups of people today who are often more egalitarian and in some respects even more morally advanced than we are – a "we" that assumes that no one from these groups of people is reading these lines).

There are no different moral universes. Moral facts have a self-affirming quality; they announce themselves – in the form of conscience or in the framework of our life experience.

In contrast to knowable and unknowable facts [*Tatsachen*] are what we can call *known* **facts** [*Fakten*]. When we speak of known facts with respect to the pandemic or climate change, we mean something that we know or for which we have sufficient robust grounds of knowledge to claim that we know. For example, it is a known fact that available vaccines can reduce the burden of disease in a population, just as it is a known fact that the virus attaches itself to our cells via its spike proteins and can lead to fatal disease through its replication. Likewise, there are a variety of known facts about climate change, including the very general known fact that it is fueled by excessive emissions and an unsustainable economy of consumerism and turbo-charged capitalism. In this sense, there are also known moral facts, to which belongs the prohibition of incest, prohibition of theft out of greed, or (positively) the commandment to save people from drowning if this does not put you in danger. The list of known moral facts can be extended indefinitely. We all have a huge range of moral knowledge that is more or less well thought-out ethically. Even Kant, who, as is well known, is seldom uncomplicated, emphasizes that "human reason can easily be brought to a high degree of correctness and accomplishment, even in the most common understanding,"[52] which is why he sees his ethics as a project to find the most comprehensive principle of moral knowledge that people have at their disposal long before they engage in philosophical analysis.

I do not mention this here to take Kant's ethics at face value. After all, unlike Kant, I do not believe that the moral facts can be derived from a supreme principle. In my opinion, however, Kant is quite right when he points out that practical knowledge, which he calls "wisdom," "which otherwise consists more in conduct than in knowledge,"[53] is a pre-theoretical source of ethical knowledge. We know much more about what moral facts exist today than we did in the past because of long historical training in everyday ethical thought and action. Ethics is making progress. Moral progress exists.

The fact that no one in democratic constitutional states today seriously doubts that women should have the right to vote is progress, as is the growing awareness of the forces of psychological violence leading us to scrutinize online harassment, sexual harassment, and the many other everyday forms of discrimination in order to find a morally better way of dealing with one another. But there is also moral regression. This means there is not simply a singular story of progress but multiple threads of a complex web. Moral regression includes the episode of the Trump presidency, fortunately over for now, and in particular the unleashing of industrial forces that has led to the destruction of our environment, which we have to contend with. In this respect, those of us in industrialized nations today are morally more regressive than the ancient Greeks, even if they were morally wrong in other respects (such as slavery). And the fact that a terrible war of aggression is currently raging in Ukraine is also a case of great moral regression, driven by an unscrupulous dictator whose entire work for decades has been clearly and unquestionably aimed at reversing moral progress. There is no sole, singular story of progress and certainly not one at the end of which stands the supposedly moral so-called West. This is why it is absolutely right (and morally progressive) that the realization is slowly spreading that we must all take seriously the indigenous knowledge that has allowed our ancestors (as well as many people whose way of life has not yet been penetrated by consumer capitalism) to live sustainably on our planet for many thousands of years. Indigenous knowledge is an important source of wisdom, which does not of course mean that groups of people we classify as "indigenous" are automatically morally correct in all matters. For that is not true of any group of human beings.

Not-Knowing

What a person knows from her own perspective can be described as **subjective knowledge**. Each of us has subjective knowledge simply because we have an individual perspective. There is knowledge from the point of view of the first person, such as my knowledge of how I feel at the moment. From this we can distinguish objective human knowledge, i.e. the sort of knowledge we can look up in textbooks and manuals, dictionaries and encyclopedias. **Objective knowledge** consists of those facts that can be known by everyone because someone has established them in one way or another. Objective knowledge is communicable in such a way that it does not matter what subjective states a person went through in order to acquire it. We learned at some point that Hamburg is north of Munich. We know this. When you learn it, you feel one way or the other and have ideas about Hamburg and Munich that others do not. But all this does not matter with respect to the fact that Hamburg is north of Munich.

It belongs to objective knowledge that no one is omniscient, not even humanity as a whole. The attempt to draw up a complete epistemological (i.e. knowledge-theoretical) map showing what we objectively know and what we do not know is bound to fail.[54]

We can never fully couple knowledge and reality. Reality, i.e. the realm of what could possibly be known, goes unknowably far beyond what we can know (subjectively and objectively). This is the **basic thesis of epistemological realism.**

Accordingly, being a realist does not mean holding a naïve view of human cognitive possibilities, as some believe. On the contrary, it means recognizing that, in principle, we cannot know everything that can be known, so that knowledge and reality cannot be completely identical under any circumstances. Our knowledge is and remains piecemeal.

But it does not follow that we should throw the baby out with the bathwater (which is never a good idea). Just because we do not know everything does not mean that we know nothing or even very little. After all, in the first place, we know that we don't know everything.

And, in the second place, we know quite a lot beyond that. We even live in a knowledge-based society that can afford to create new possibilities of not-knowing through its knowledge. On the basis of our impressive technological knowledge, we have built machines and connected them to an internet that has now maneuvered us into the paradoxical position of seducing many into believing things that are false through fake news, misinformation and subtle manipulation. Just think of the American debate between hardline creationists and evolutionary biologists, who now know that the forms of living things can be biologically explained without our having to assume that God created all living things spontaneously at the beginning of creation (which is still echoed in the word "creatures" used for non-human living things, i.e. the 'animals'). Hard-core creationism is false, but it is a widespread belief that is now spreading rapidly as an infodemic through what the Oxford internet researcher Philip N. Howard has called the "lie machines" of the internet.[55]

It is a remarkable anthropological fact that we humans can become convinced that we do not know something that we actually know. We can believe that we know less than we know. This is because objective knowledge is not necessarily associated with a particularly strong power of persuasion.

Against the backdrop of fake news, digital propaganda and manipulation, the proliferation of deep fakes (i.e. videos and photos that appear to depict scenes that were digitally programmed and never took place), social networks, and so on, a new debate has emerged in epistemology about epistemic vices. These include excessive news consumption and social media use.[56] Whoever follows too closely the countless news tickers on daily political topics (which even the serious quality media now offer in their online outlets, often paired with clickbait that links to articles behind a paywall), will certainly very quickly become extremely confused on account of the accumulation of contradictory and emotionally charged information.

Paradoxically, our technological knowledge enables new forms of not-knowing, some of which are socially dangerous. Moreover, we must bear in mind that we do not know on the whole whether we are on a straightforward path of scientific and technological progress. What could be progressive about the fact that we are still under the misapprehension that nature is a collection of lifeless raw materials that we can use at will for our consumer needs? This error is based on a false philosophy of nature that continues to follow the model of the modern view of nature as a great machine in which everything is mindless and unconscious except for the actions of human beings. Such a philosophy of nature is wrong at the very least because we do not know which life forms are conscious and intelligent. Are bacteria already conscious, living in intelligent swarms? If intelligence is seen as the ability to solve problems, or even as the ability of a living being to adapt to its environment, unicellular organisms are more intelligent than humans. And if a certain neural complexity is the prerequisite for consciousness, at what point is the threshold of complexity reached? Are insects conscious, or only some bird species and mammals? What about plants or forests, assuming we find patterns of complex organization in their networks that are comparable to the complexity of a brain? This brings us back to the anthropocentric worldview from the first part of this book. Plants seem to be further away from us, which is why we regard them as less intelligent, even less valuable. But how do we know that we inflict less pain on broccoli when we eat it than on insects? We should probably even be prepared to see the nature of plants not always in comparison to 'animals.' And how do we know that in a human being there is only one consciousness that is associated with cortical brain regions? The nervous system in the gut influences neural processes in the brain region. Perhaps it is for its part conscious and intelligent. If so, we would require a completely different form of ethics.

We are far from bringing our third-personal, objective knowledge of nature into congruence with our subjective, first-personal knowledge. Many even consider the problem to be unsolvable, because the scientific language (of mathematical modeling and experimental testing) is not suitable in principle adequately to capture the subjective side of reality.

One of the most prominent representatives of the thesis that we cannot fully grasp nature objectively due to its inner, subjective side is the physiologist Emil Heinrich Du Bois-Reymond, who in Leipzig on August 14, 1872, gave a lecture that is exemplary to this day: "Limits of Our Knowledge of Nature." There he says in a famous passage, just as relevant today as back then:

> It is absolutely and forever inconceivable that a number of carbon, hydrogen, nitrogen, oxygen, etc., atoms should not be indifferent as to their own position and motion, past, present, or future. It is utterly inconceivable how consciousness should result from their joint action.[57]

The third-person, objective knowledge of physics, chemistry, biochemistry and neuroscience is in principle incapable of comprehending "mental phenomena ... from their material conditions."[58] According to Du Bois-Reymond, it is equally impossible to explain the material-energetic structure and the forces that we find in the universe from elementary principles, principles that would be even more fundamental and plausible than what is found by physics. At some point physics runs up against *brute fact*, i.e. the fact the measurements are just the way they are. How physical reality is cannot be derived from any principle that would make its nature more understandable. This is already built into the fact that we can know only one physical reality, namely the one to which our measuring instruments – both our physical measuring instruments and our perceptual organs – have access. At the same time we know today much more exactly than in the 1870s that this area of study, namely the universe, is limited in many ways. The assumption that the natural sciences are approaching a complete worldview is completely untenable for logical, for philosophical and also for scientific reasons.

The standard and still widespread scientific worldview, according to which we supposedly know that we are in a gigantic material-energetic system whose successive states explain absolutely everything that actually takes place, is a piece of ultimately ludicrous speculation and untenable epistemic hubris. In modernity, this hubristic worldview has brought us to the situation of imminent self-extinction.

Towards an Ethics of Not-Knowing

Nature demands the utmost respect from us. The more precisely we explore it by improving our instruments, models and theories, the clearer it becomes that its complexity exceeds anything we can imagine. But nature is by no means only "outside of us." It is not only the object of scientific cognition or perceptual observation; it is, rather, also "in us." For we are animals in which mind and matter meet and are merged in a unity we do not really understand. In order adequately to understand the unity of mind and matter we embody would require large-scale cooperation among the various forms of knowledge with the goal of finally understanding ourselves as human beings. This presupposes that the natural sciences, the technical sciences, the humanities and the social sciences work on common questions, without falling into the mistake that all knowledge can ultimately be reduced to an epistemically privileged layer (whether this is mathematically formulated knowledge or the qualitatively accessible experience of historically situated subjects, which the humanities explore).

Wilhelm von Humboldt, one of the inventors of the modern research university, and many other thinkers of his time, who asked themselves how universities (and thus academic knowledge) should relate to social progress, sought to assemble human knowledge in terms of the question of who or what the human being is. Of course, much of what was written and thought on this subject at that time is hopelessly antiquated. Nevertheless, the basic idea of the Humboldt ideal of education, which underlies many university reforms around 1800, is correct in principle. We know today that the natural and social phenomena that entangle humanity in unprecedented crises in the age of modernity can only admit of potential solutions if we recognize the irreducible complexity and thus the interconnectedness of these phenomena. Therefore, we do not need science as a complexity reduction. Rather the opposite. The goal of science must be the adequate comprehension of complexity. That means the mereological nesting of fields of sense which reaches self-knowledge in human life. Respecting the limits of knowledge is a crucial part of this self-knowledge. An ethics of not-knowing does not advocate for superstition or "alternative facts." Rather, it begins by recognizing that we cannot get rid of the human being. We are and remain vulnerable, finite and limited living beings. But we can also become aware of this status. Thanks to our embodiment, i.e. thanks to the fact that we are animals, the sensual experience of nature, art, sexuality and a rich emotional life are available to us. These we can enjoy and cultivate as long as we live. As we become aware of nature on the level of the mind, nature partially reaches a surprising form of self-knowledge.

Ethics is also part of this self-knowledge. For in ethics we articulate in a conscious and explicit way which values we have and which of these ideas are right or wrong.

Now is not the first time we have lived in uncertain times. To live means to live in uncertain times, for all individual life (and probably also all life in general) in our universe will come to an end one day. As far as we know today, life is an obsolete model that will come to an end at the latest with the extinction of our sun in our corner of the Milky Way. Life is a risk, whereby we all always lose everything in the end.

Part of the ethics of not-knowing is that we overcome the misguided fantasy that, through scientific, technological and (above all) medical progress, we could create a kind of paradise on Earth that would give us a life as close to eternal as possible. Don't get me wrong: medical progress should be welcomed. But it must be embedded in ethical reflection and thus linked to conditions of moral progress.

*The starting point of the ethics of not-knowing is **epistemic modesty**, a knowledge that our knowledge claims concerning nature, as well as our concrete value concepts and ethical judgments, are prone to error and in need of correction. This epistemic modesty is grounded in our respect for the other, which always remains alien to us.*

The ethics of not-knowing is therefore based on knowing. It makes its own claims to knowledge, which are derived from epistemology, the theory of science, natural philosophy and philosophical ethics. In general, as mentioned, ethics is a sub-discipline of philosophy. It is a mistake to believe that it belongs to biology or to theology, or that it is a matter of building social consensus. Ethics as such is secular and humanistic; it arises from the self-knowledge of the minded living being called "human being," the being which can recognize itself as an animal that does not want to be one.

The knowledge claims of philosophy are not isolated or even superior to other knowledge claims. On the contrary, philosophy can only accomplish its work in constant conversation with the other sciences in order to apply its own methods with the aim of getting our humanity into view. This goal never pursues purely theoretical cognitive interests. Instead, ever since the founding acts of philosophy in Africa, Asia, Greece, and so on – the exact prehistory no one knows – it has been linked to recognizing the contours of the idea of the good. We humans are the guardians of this idea. It is our moral mission, the meaning of life. We can only hope that we will accept it and thus do justice to the name of our species, our capacity for wisdom.

Acknowledgments

This book owes its existence to a number of individuals and institutions. First and foremost, the New Institute in Hamburg supported the work through a fellowship in 2021–2. There I had the opportunity to discuss some of the main theses in the program "Foundations of Value and Values," as well as with other fellows and colleagues. In alphabetical order, I am particularly grateful to Harald Atmanspacher, Georg Diez, George Ellis, Christoph Gottschalk, Christoph Horn, Anna Katsman, Wilhelm Krull, Anna Luisa Lippold, Corine Pelluchon, Dennis Snower and Ingo Venzke, with whom I was able to discuss various aspects of this project.

In the last few years, the constant exchange with Daniel Kehlmann has been intellectually and personally very enriching for me. Daniel already pointed out to me in the fall of 2019 that we must always view the moral burden of the ecological crisis from the perspective of future generations and measure our conception of ourselves as minded beings against our imagined futures. We must look back at ourselves from an imagined future and, in this way, set normatively desirable courses for the present; philosophy and ethics are futuristic. Daniel's hints are informed by his outstanding literary imagination, which goes far beyond the usual apocalyptic patterns of thought in which ecological discourse sometimes remains trapped. It is necessary to think post-apocalyptically, without assuming that the future will only come about once an actually avoidable apocalypse has occurred. Such a perspective opens up a world of imagination that is both realistic and utopian, something Juli Zeh convinced me of in an illuminating conversation.

I thank my university and especially its rector, Prof. Dr Dr h. c. Michael Hoch, not only for granting me a sabbatical to carry out the fellowship but especially for the main idea of the book, i.e. to shed light on the animal character of human beings on different scales. Michael Hoch has long been insisting that we need to take microorganisms seriously in our conception of human life because they are essential drivers of the evolution of living things – a theme he has brought to my attention during our encounters for a decade now. Anyone who still doubts the colossal importance of microorganisms has been on a different planet in the recent pandemic years.

I would like to thank my colleagues from the life sciences in Bonn, with whom I was able to discuss various aspects of this book – from virology and neurobiology to the cosmological framework of the emergence of life. Not being a life scientist myself, the excellent environment in Bonn helped me to develop my philosophical views in encounters with information about the state-of-the-art scientific research. All misunderstandings and errors in life science, which may nevertheless be found in this book, are exclusively my own. I would like to thank especially, again in alphabetical order, Heinz Beck, Waldemar Kolanus, Ulf-G. Meissner, Joachim Schultze, Hendrik Streeck, Christiane Woopen and Andreas Zimmer, with whom I exchanged views on various occasions (most recently at an inspiring symposium at the Center for Science and Thought). This was made possible in an optimal way by the transdisciplinary research areas at the University of Bonn.

Further thanks go to Gert Scobel for the many intensive discussions on topics at the interface of natural sciences and philosophy, as well as on questions of an ethics of transformation, which always revolved around the problem of complexity, which is at the center of his reflections.

As always, special thanks go to my team at the Department of Epistemology, Modern and Contemporary Philosophy, who were actively involved with criticism, comments and corrections during the preparation of the manuscript. Without such a lively, intellectually honest and inspiring context, it would not have been possible to complete the book. Again, any errors are my fault alone. I would like to thank Philipp Bohlen, Alex Englander, Charlotte Gauvry, Sergio Genovesi, Joline Kretschmer, Leila Mahmutovic, Laura Michler, Jens Rometsch and Jan Voosholz.

Most recently, the intensive discussions on philosophy, consciousness, life and neuroscience at the Fundación Tatiana Pérez de Guzmán el Bueno retreat in Cáceres and Madrid in November 2021 were particularly helpful. I thank Matthew Cobb, Thomas Fuchs, Alva Noë and Georg Northoff for the references they contributed to the neurophilosophical deep structure of the reflections published here. The hospitality of the foundation and the visionary idea of delineating a new neurophilosophical framework that constitutively considers research ethics are pathbreaking.

At this point I would also like to thank Juan Ezequiel Morales, who makes unforgettable intercultural encounters possible in Gran Canaria, which are reflected in this book insofar as he has been assuring me for years that living things form a large, predominantly invisible fabric that deserves the name *grande ser* ("great being").

It almost goes without saying that I owe my family everything that has value in the depths of a book and a life, and should therefore be expressly emphasized here with an acknowledgment to my loved ones. Thinking takes place not in solitude but in a group that sometimes has patiently to endure the fact that writing only succeeds when you retreat into solitude. Thank you for your support, which means everything to me, and for always interrupting my writing and reading with tapping, laughing, jumping – in short, with real life.

Finally, as always, I would like to thank Julika Jänicke, publishing director of non-fiction at Ullstein, who has read and commented on the book at various stages. She also inspired it because she herself thinks intensively about the topics of life and science and the connection between ethics, politics and being an animal. Last but not least, I would like to thank Uta Rüenauver for her competent editing; she helped tweak the manuscript along the home stretch.

Glossary

Amphibian theory: The amphibian theory says that humans lead two forms of life (from the Greek *amphi-bios* = double-lived). On the one hand, we are animals like all other animals; on the other, we live the higher life of reason.

Animalism: The thesis that we humans are animals of a certain species. This species is referred to as *Homo sapiens*. In less technical terms, each of us is a human animal. According to animalism, a human being is fully and completely an animal.

Anthropocene: Term for the geological era in which humans have become the decisive geological factor, which is measurable, among other things, through our globally noticeable impact on climate change.

Anthropological difference: The anthropological difference lies in the fact that humans are different from other living beings.

Anthropological dualism: The thesis that a human being consists of two parts, an unreasonable, animal part and a specifically human part, usually regarded as superior.

Anthropology: The science of human self-investigation.

Anthrozoology: The science of human self-investigation insofar as we see ourselves as animals.

Autonomy: The capacity of a living being, paradigmatically a human being, to give itself a law, i.e. to measure its own actions intentionally against a standard which one must oneself recognize in order for it to apply.

Autopoiesis: Self-generation and self-maintenance of a living system.

Basic idea of liberal pluralism: The goal and meaning of our earthly life is that we delineate margins of social freedom which allow every human being to find meaning in life, whatever this may then consist of in the individual.

Basic thesis of epistemological realism: We can never fully couple knowledge and reality. Reality, i.e. the realm of what could possibly be known, goes unknowably far beyond that which we can know (subjectively and objectively).

Being-for-itself (of the mind): The being-for-itself of the mind consists in the fact that the mind exists only by relating to itself in one way or another (such as in the form of consciousness). The existence of the mind does not remain unnoticed, because it exists as noticing itself. The mind essentially reveals itself to the mind; it cannot remain hidden.

Being-in-and-for-itself: Being-in-and-for-itself is the entanglement of nature and mind that manifests itself in humans as animals. Minded life is therefore, on the one hand, manifest (for-itself) – it expresses itself – but, on the other hand, it is not completely knowable due to its natural foundations and is therefore in-itself.

Being-in-itself (of nature): Being-in-itself consists in the fact that nature is as we find it, without its being directed towards us or being interested in any other way in being perceived by us and grasped in the form of natural scientific theory.

Biological continuity: The continuity of a person over different periods of time. It is seen in the biological unity of the person's organism.

Biological relativity: Which part has an influence on which whole and vice versa depends on the perspective associated with an organ in the organism itself, the environment of which constitutes an environment for it in exactly the same way as the reality around us is experienced by us as a non-self or an external world.

Biologism: The thesis that the meaning of human life in all its social, historical, cultural and institutional manifestations and therewith our social freedom can ultimately be traced back to the fact that we are animals whose life form can be completely deciphered by biology.

Böckenförde dictum: "The liberal, secularized state is nourished by presuppositions that it cannot itself guarantee."

Cartesian dualism: Assumption, dating back to René Descartes, that there are two substances in nature, i.e. two kinds of things – the extended, material bodies and the immaterial souls or thinking substances – which somehow causally interact.

Category mistake: An inadmissible transfer of forms of thought and facts that apply in one area to another area in which they do not apply in the same way.

Complexity: The fact that our analysis and explanation of a system dissolves or changes the very system that we want to analyze and explain by breaking it down into subsystems. Therefore, complex systems can never be completely analyzed and explained, which is why predictions of their behavior are ultimately impossible.

Continuity thesis: The continuity thesis states that humans are animals in the same sense as all other animals and that every characteristic that distinguishes us from other animals can ultimately also be observed elsewhere in the animal kingdom, albeit in a different form. Accordingly, humans are not different from other animals in principle but only in degree.

Creationism: The thesis that all life processes are somehow directly caused and controlled by God, so that modern evolutionary biology is based on false premises.

Criterion of reproductive isolation: The explanation of the unity of a species by the fact that it can only reproduce with members of the same species.

Derrida's animal-philosophical insight: There is nothing in reality that corresponds to our confused concept of the animal.

Discontinuity thesis: Humans differ from other animals in principle in a number of characteristics.

Ecocentrism: The view that the environment is the real reason for urgently needed human action, even for a complete transformation of our societies.

Ecology: The study of processes of systemic adaptation and system production by living things on a scale that takes into account life forms and their causal environment. This is a mode of thought in the environmental sciences.

Environment: The field of sense within which a life form can reproduce itself over several generations.

Epistemic: Pertaining to our knowledge, from Greek *epistêmê* = knowledge.

Epistemic modesty: The starting point of an ethics of not-knowing; the knowledge that our knowledge claims regarding nature as well as our concrete values and ethical judgments are fallible and corrigible.

Epistemic moral realism: The thesis that we can recognize moral reality and to a certain extent have always recognized it correctly.

Epistemically objective: Something (including a subjective state) is epistemically objective if one can have opinions about it that can be true or false.

Epistemological foundationalism: The assumption that there is an immediately given layer of elementary, conscious experiences and sense data that cannot be doubted any further. (It has been refuted since an influential book by the American philosopher Wilfrid Sellars.)

Ethics: The sub-discipline of philosophy that deals with the question of what we should do or refrain from doing – just insofar as we are all human beings.

Existential self-determination: A view of ourselves that we adopt with reference to our image of humanity, thus defining ourselves as human beings at our core.

Existential transcendence: Our own self-determination derives its meaning from something that transcends it and in which it is embedded.

Extreme fictionalist thesis: Basically, all entities and relations that play a role in the natural sciences (numbers, forces, causality, space, time, etc.) are fictions that the human mind constructs in order to form a simplified but distorted picture of the infinitely complex and chaotic experiential reality that results from the stimulation of our senses.

Fact [*Tatsache*]: A true answer to a meaningful question.

Facts [*Fakten*]: Known facts [*Tatsachen*].

Falsificationism: The theory of science which states that the truth of scientific statements can never be proven directly, only their non-falsity.

First-person standpoint (subjectivity): The point of view from which we know something about ourselves that we can only experience subjectively.

Form of survival: The organic life that the life sciences successfully research, frequently for the benefit of humanity.

Fundamental anthropological principle, first: Our concept 'animal' is always designed in view of humanity, from whom we take our starting point in our investigation of the animal kingdom. What appears to us as animal, we consciously or unconsciously take from our own nature and then project more or less uncritically onto the other forms of life. In this way we erect a distorting mirror in which we no longer recognize ourselves and the other living beings correctly. In place of the question of how humanity relates to the animal, we should therefore ask how humans determine themselves as animal.

Fundamental anthropological principle, second: The human being is the animal that does not want to be one. By establishing our 'animality,' we are at the same time anxious to strip it off and to identify ourselves with that part of ourselves by which we differ from the other animals.

Fundamental anthropological principle, third: The human being is the only animal known to us, while the other forms of life are utterly alien to us and in reality should not be defined as animals by human standards.

Good life: The good life consists of understanding ourselves as living beings capable of higher, universal morality, making us responsible actors in the kingdom of ends.

Holism: The doctrine of the whole (from the Greek *to holon*). Here this means the assumption that an organism acts as a whole on its parts.

Humanism: Ethics, as reflection on what we should do or refrain from doing, is *anthropogenic*, i.e. it arises from the human form of life, without therefore being *anthropocentric*, i.e. directed only at humanity. The human being is the starting point of ethics.

Hygienism: The reduction of humans to animals to be classified as healthy or sick.

Hyper-complexity: Although mind and nature are interwoven (complex), they are nevertheless categorically distinct from each other and therefore hyper-complex.

Idea of the good: The assumption that there is something that sheds an evaluative light on human actions even independently of how the actors classify them. The idea of the good is the reason why we experience reality as having value.

Identity trap: This fallacy deduces, from the fact that mind occurs (so far as we know) only when certain natural, organic conditions are fulfilled, that these conditions are not only necessary but also sufficient presuppositions for mind. Then the mind would be identical with exactly these conditions.

Image of humanity: A self-determination of man, i.e. a concrete understanding of what being human actually means.

Immaterialism: Thesis according to which we have at our disposal a thinking, immaterial substance.

Linguistic turn: All thinking is ultimately an expression of language or dealing with signs, so that linguistics becomes a theory of thinking and a method of philosophy.

Living body: The living body [*Leib*] (in contrast to the physical body [*Körper*]) is our organization as it is experienced from the perspective of the first person, i.e. subjectively.

Lockeanism: Doctrine according to which the unity of a person over different times consists in the fact that they consider themselves to be the same entity.

Logic: The basic discipline of philosophy whose object is rational thinking itself. It is concerned with the question of how we think correctly (i.e. rationally), i.e. how we ought to think, and its form of implementation is thinking about thinking.

Makropulos thought experiment: A thought experiment that examines whether an immortal earthly life makes sense at all or whether it would not, rather, be a tragedy for those concerned.

Meaning of life: According to the original Enlightenment and religious idea, the meaning of life is that we are destined in our lives to work together, as a human community of destiny, to make the best possible life feasible for all people.

Mechanism: Mechanism as a form of thinking consists in understanding a system of elements (such as cells), arranged and interrelated in a certain way, from the simplest possible interaction between its parts. The sum of the individual parts is therefore no more than these individual parts in their interaction.

Mereology: The study of wholes and their parts.

Metaphysical scientism: The doctrine that objectivity can only be achieved by the stripping off of our subjectivity and developing a scientific view of the world in the form of studies, measurement data and mathematical models that capture reality as it actually is.

Mind: The capacity to lead one's life in the light of a conception of who or what one is.

Modern nihilism: The thesis that nature itself has no meaning or significance. According to this view, nature is simply a material-energetic total structure in which living beings have at some point emerged. These beings evaluate their environment, which is the origin of values.

Moral facts: A true answer to an ethically relevant question that can be true or false.

Moral realism: Assumption that there are objectively correct convictions about what we should do or refrain from doing, just insofar as we are human.

Naturalism: Doctrine according to which the world or reality as a whole can be completely explained and controlled by understanding humanity as a natural phenomenon, as an element of a gigantic material-energetic system in which systems of varying complexity interact.

Nature: The target system of scientific research, specifically physics, chemistry and biology.

Neo-Darwinism: The combination of Gregor Mendel's theory of heredity with modern genetics (which has finally proven the existence of genes), i.e. Darwinism plus genetics.

New Realism: View according to which our experience of reality, the mind or subjectivity belongs to reality on an equal footing with that which we justifiably classify as independent of us. Therefore objectivity does not consist paradigmatically in the fact that we abstract from subjectivity. The objective includes the subjective (and vice versa): objectivity and subjectivity are mutually contained within each other.

Normativity: The fact that our thinking and acting are classified as correct or incorrect.

Objection of two lives: If the human being consists of spirit and nature, then he leads two lives at once, the life of a specimen of the animal species *Homo sapiens* and the life of a thinking person who is never completely identical with his body. The objection classifies this assumption as a problem.

Objective knowledge: Knowledge that consists of those facts that can be known by everyone because someone has established them in one way or another.

Ontic: Pertaining to being (from Greek *to on* = being). It is distinguished from "epistemic," i.e. concerning our knowledge claims regarding being.

Ontically subjective: A fact is ontically subjective if it only exists because subjective states (feelings, opinions, sensory impressions, etc.) are involved in it.

Ontological fact realism: The assumption that the realm of facts goes indeterminately far beyond what we can express linguistically.

Organism: A complex system of organs such that the more or less delimited whole of this system has a partial influence on how its organs develop.

Organs: Instruments (such as our organs of perception – eyes, ears, skin, etc.) shaped by evolutionary processes. The instruments capture a partially independent reality on a level assigned to them.

Person: A self that has certain rights and duties. A person is defined by social and moral relationships and is not absorbed into the structure of the material-energetic processes of self-preservation through which an animal maintains itself as an organism against environmental pressure.

Phenomenology: Phenomenology examines the way in which reality appears to us.

Philosophical anthropology: An area of knowledge that deals with the question of who or what a human being is. Every concrete answer to this main question of anthropology provides a conception of humanity.

Physical body: The physical body [*Körper*] (in contrast to the living body [*Leib*]) is our organism as it appears as an object from the scientific point of view of the third person (such as a doctor).

Posthumanism: The assumption that any distinction between humans and nature is ultimately obsolete and owes its existence to outdated ways of thinking that set humans apart from their environment and thus see them directly or indirectly as the crown of creation.

Principal thesis of neo-existentialism: People are free insofar as they have to create an image of themselves in order to be someone in the first place.

Privative zoology (from the Latin *privare* = to deprive) (after Pelluchon): A theory that understands non-human animals as animals in the same sense as we understand ourselves as animals but recognizes a deficiency in the other animals because they lack precisely that which distinguishes us from them.

Problem of speciation: The problem of formation of species.

Problem of temporal personal identity: This problem lies in the fact that it is not easy to say in what respect (if at all) I am roughly the same as I was in the past.

Projection structure: First, humans determine their own animality and look for characteristics that they have in common with other animals. This then gives rise to the concept of the mere animal (the animal in general). In the next act, being an animal is then applied to humans themselves in order either to domesticate them through culture and civilization or to identify with them normatively.

Projection thesis: We create the idea of a unified animal kingdom on the basis of a human, all-too-human representation of the animal, which can be seen through as a largely indirect expression of a disturbed relationship between ourselves and our own 'animality.'

Psychological continuity (= Lockeanism): According to this approach, I am the same as I was in the past because there is a mental connection between my present self and my former self.

Radical alterity: The other in us and the others in general (other human beings, nature, other living beings, God and the gods, if one believes in them, and so on) are others precisely in that they have dimensions which we do not know. If the others in general and the other as such were an open book for us, completely knowable, they would not be really other, but predictable repetitions of the same.

Radical autonomy: We recognize our human capacity to lead our lives in the light of an idea of who we are and who we want to be as a source of value; thus we always measure our individual autonomy against the autonomy of others. Autonomy thus becomes the measure of itself, because autonomy is measured against autonomy.

Raven paradox: If we have observed only black ravens for a long time, we can formulate the statement "All ravens are black." By doing this, we have ruled out the possibility of any raven being any other color. In this way, we have developed a fairly simple theory. This theory is falsified, thrown overboard, by someone showing us a single raven that is not black. That all ravens are black is therefore not confirmed definitively by any series of observations, no matter how good they may be, but is always valid only provisionally, i.e. as long as no non-black raven has been found.

Reductionism: The reduction of a whole to its parts.

Saying of Anaximander: "And the things out of which birth comes about for beings, into these too their destruction happens, according to obligation: for they pay the penalty (*dike*) and retribution (*tisis*) to each other for their injustice (*adikia*) according to the order of time[.]"

Scientific fictionalism: The view that every scientific theory is a fiction, i.e. a mixture of more or less consciously accepted half-truths, unrecognized falsehoods and non-falsities, but also of truths and precise measurements.

Scientism: Scientism is the idea that there is a singular, ultimately omniscient science that will prevail socio-economically through experimentation, rational theory construction and expertise, finally freeing mankind from all evils.

Social freedom: We can expand our autonomy by cooperating with the autonomy of others. Freedom consists of doing something together with others that you cannot do alone.

Social projection: An idea of the otherness of some given (factually existing or even altogether imagined) group, in which the initial group uses its self-image to measure the foreign group against it.

Sociologism: The thesis that sociology can only ever describe social reality in a value-neutral way and that, in reality, there are no objective values, only the positings of subjects.

Solutionism: The idea that technology can solve any problem in no time at all, that all you have to do is invest huge sums in it if necessary.

Soteriology: Doctrine of salvation.

Speciesist prejudice: The prejudice of one species against another or all others that it is far superior to other limited species on account of its peculiarities (particularly common in humans).

Specific difference: What distinguishes a species from all other species within the same genus.

Stereotype: Simplified and ultimately misguided explanation for the fact that others are different.

Strong anthropic principle: This principle states that the universe necessarily runs towards its self-knowledge in cognizing living beings such as humans.

Strong anti-realism according to Korsgaard: A view according to which moral standards are nothing more than the standards of behavior that we can all want people to adhere to. Accordingly, not everyone can agree with this behavior because it is morally right, but it is morally right *because* everyone can agree with it.

Subjective knowledge: Something a person knows from their own perspective.

Technology [*Technologie*]: An attitude towards technology [*Technik*], natural science and economy.

Third-person standpoint: The point of view from which we gain objective knowledge about objects of natural and social scientific research. This standpoint is equally accessible to every subject.

Transcendent normativity: The idea that, even if an entire society had a unanimous opinion regarding certain actions, there would still be an independent standard of evaluation that classifies their actions and thoughts as correct or incorrect. This evaluative standard therefore exceeds the evaluation guidelines that this society has implicitly or explicitly given itself.

Transhumanism: The view that humans should be overcome in favor of a higher form of life, possibly produced by technology.

Universalism: The idea that what we should do or refrain from doing for moral reasons is valid for all people, i.e. it is cross-cultural and in principle also recognizable for all people in their respective contexts.

Universe: The field of sense that we capture with our mathematical-physical models.

Vitalism: The thesis that living beings, which we classify as organisms, form wholes that are more than the sum of their parts.

Weak anthropic principle: The observed values of all physical and cosmological quantities are not equally probable. Rather, they assume values that are constrained both by the requirement that there be places where carbon-based life can arise and by the further requirements associated with the universe being old enough for this to have already occurred.

Wise words of Socrates: *"I have become aware that I know nothing"* (emautôi synêidê oudén epistamenôi).

Wolf's love thesis: The thesis that the meaning we find in life is that we can love something or someone.

Zoo theory: The idea that, just as in a zoo, we put the life forms we encounter into enclosures at our discretion and according to our categories. These include zoological classifications.

Notes

Introduction

1 An earlier version of the first pages of this introduction first appeared in the *Neue Zürcher Zeitung* on 9 July 2021. On the concept of modernity and the diagnosis of the current complex crisis scenarios as "late modern," see Andreas Reckwitz and Hartmut Rosa, *Spätmoderne in der Krise: Was leistet die Gesellschaftstheorie* (Berlin: Suhrkamp, 2021).

2 See in particular Corine Pelluchon, *Das Zeitalter des Lebendigen: Eine neue Philosophie der Aufklärung* (Darmstadt: wbg Academic, 2021).

3 In the following, I will not use grammatical genders throughout, but will vary them. In doing so, I am not expressing any opinion on the extent to which gender-neutral language is an ethically appropriate way to counteract the various forms of discrimination to which people are exposed on the basis of their gender and other factors of their humanity. Where it has been chosen, the choice of the generic masculine is only grammatically motivated, without this motivation implying that I want others or even the "English language" to adhere to this convention. In the examples and where more recent gender conventions are used, I try to balance out the use of the generic masculine (and feminine, which I also use from time to time) in order to counteract stereotypes that implicitly or even explicitly classify certain activities, professions, etc., as 'male' or 'female,' not least of which is the word "philosopher," whose genus has been associated with 'men' for far too long.

4 In the book, single quotation marks signal a distancing. When I write 'male,' 'female,' 'animal' or 'society,' for example, I am distancing myself from the idea that 'out there,' 'in reality,' there is something that many people associate with these terms.

5 The expressions in bold are cornerstones of my thought process in this book. You can look up their meaning or definition in the glossary.

6 See Steven Pinker, *Rationality: What It Is. Why It Seems Scarce. Why It Matters* (London and New York: Random House, 2021).

7 Daniel Kahneman, *Thinking Fast and Slow* (New York: Farrar, Straus & Giroux, 2021), and also Daniel Kahneman, Olivier Sibony and Cass R. Sunstein, *Noise: A Flaw in Human Judgment* (Boston: Little, Brown, 2022).

8 Lorraine Daston and Peter Galison, *Objectivity* (New York: Zone Books, 2010).

9 See Tyson Yunkaporta, *Sand Talk: How Indigenous Thinking Can Save the World* (New York: Harper, 2020). See also similar reflections in Bayo Akomolafe, *These Wilds beyond Our Fences: Letters to My Daughter on Humanity's Search for Home* (Berkeley, CA: North Atlantic Books, 2017).

10 See Markus Gabriel, *I am Not a Brain: Philosophy of Mind for the Twenty-First Century* (Cambridge: Polity, 2017), as well as Markus Gabriel, *Neo-Existentialism* (Cambridge: Polity, 2018), with contributions by Jocelyn Benoist, Andrea Kern, Jocelyn Maclure and Charles Taylor.

11 A further note: passages in italic type contain key statements.

12 And it is on this very basis, as described in Reckwitz and Rosa, *Spätmoderne in der Krise*, and proceeding from the anthropology of the Canadian philosopher Charles Taylor, that the authors derive a normative framework of social theory, thus avoiding **sociologism**, i.e. the thesis that sociology can only ever describe social, evaluative reality in a value-free way and has thus discovered that, in reality, there are no objective values but only the positings of subjects. "I would exclude these two phenomena, that interpretation must be interpreted and that interpretation implies values, from historicization and declare them to be universal: acting human beings are always and everywhere self-interpreting beings who orient themselves on the basis of a map of strong valuations" (p. 278).

13 Donella Meadows, Dennis Meadows, Jørgen Randers et al., *The Limits to Growth: A Report for the Club of Rome's Project on the Predicament of Mankind* (New York: Universe Books, 1972).

14 This is the original French title (*Les Lumières à l'âge du vivant*) of Corine Pelluchon's *Das Zeitalter des Lebendigen*.

15 Yunkaporta, *Sand Talk*.

Part I: We and the Other Animals

1 *Ideas for a Philosophy of Nature*, trans. Errol E. Harris and Peter Heath. Cambridge: Cambridge University Press, 1988, p. 40. [Trans.]

2 Aristotle, *Politics* (Indianapolis: Hackett, 2017), p. 126.

3 See Eric Kandel, *The Age of Insight: The Quest to Understand the Unconscious in Art, Mind, and Brain, from Vienna 1900 to the Present* (New York: Random House, 2012).

4 Thus Freud, in a much quoted passage of his essay "A Difficulty in the Path of Psychoanalysis," in *The Standard Edition of the Complete Psychological Works of Sigmund Freud*, Vol. XVII: *1917–1919* (London: Hogarth Press, 1955), p. 143:

It is thus that psycho-analysis has sought to educate the ego. But these two discoveries – that the life of our sexual instincts cannot be wholly tamed, and that mental processes are in themselves unconscious and only reach the ego and come under its control through incomplete and untrustworthy perceptions – these two discoveries amount to a statement that *the ego is not master in its own house*. Together they represent the third blow to man's self-love, what I may call the *psychological* one. No wonder, then, that the ego does not look favourably upon psycho-analysis and obstinately refuses to believe in it.

5 Ibid., pp. 140–1: "We all know that little more than half a century ago the researches of Charles Darwin and his collaborators and forerunners put an end to this presumption on the part of man. Man is not a being different from animals or superior to them; he himself is of animal descent, being more closely related to some species and more distantly to others."

6 Freud distinguishes in many places and in different respects between being human and being an animal, even if he simultaneously claims that we are merely animals. In *Civilization and its Discontents* (New York: W. W. Norton, 1961), pp. 46–7, for example, one reads that

> It is time that we should turn our attention to the nature of this culture, the value of which is so much disputed from the point of view of happiness. Until we have learnt something by examining it for ourselves, we will not look round for formulas which express its essence in a few words. We will be content to repeat that the word culture describes the sum of the achievements and institutions which differentiate our lives from those of our animal forebears and serve two purposes, namely, that of protecting humanity against nature and of regulating the relations of human beings among themselves.

7 Relevant recent works on this animal-philosophical topic are by Markus Wild (he will forgive me the hint: *nomen est omen*). See in particular Markus Wild: *Die anthropologische Differenz: Der Geist der Tiere in der frühen Neuzeit bei Montaigne, Descartes und Hume* (Berlin and New York: De Gruyter, 2006), as well as his commendable introductions *Tierphilosophie zur Einführung* (Hamburg: Junius, 2008) and, together with Herwig Grimm, *Tierethik zur Einführung* (Hamburg: Junius, 2016).

8 See Simone de Beauvoir, *The Second Sex* (New York: Vintage, 2009), p. 466: "One is not born, but rather becomes, woman."

9 Mind you, our being an animal is not identical with the fact that our bodies exist. This elementary biological fact, as well as many other human medical facts, is largely independent of how we perceive ourselves.

10 Aristotle, *Politics*, p. 126.

11 Markus Gabriel, *Moral Progress in Dark Times* (Cambridge: Polity, 2023) and Markus Gabriel and Gert Sobel, *Zwischen Gut und Böse: Philosophie der radikalen Mitte* (Hamburg: Edition Körber, 2021).

12 Barry Schwartz, *Why We Work* (New York: Simon & Schuster, 2015), p. 14.

13 Markus Gabriel, *I am Not a Brain: Philosophy of Mind for the Twenty-First Century* (Cambridge: Polity, 2017).

14 Aristotle, *Metaphysics* (Indianapolis: Hackett, 2016), p. 113.

15 For example, the *Taschenlehrbuch Zoologie*, ed. Katharina Munk, with collaboration by Hartmut Böhm, Jutta Heidelbach, Christian Hölscher et al. (Stuttgart and New York: Georg Thieme, 2010), p. 5.

16 Daniel C. Dennett, *Darwin's Dangerous Idea: Evolution and the Meanings of Life* (London: Penguin, 1995), p. 44.

17 Ibid., p. 45.

18 For a discussion of the philosophical question of how species can actually be defined, see Marc Ereshefsky, ed., *The Units of Evolution: Essays on the Nature of Species* (Cambridge, MA: MIT Press, 1992).

19 See Joachim Bauer, *Das empathische Gen: Humanität, das Gute und die Bestimmung des Menschen* (Freiburg: Heder, 2021).

20 For an overview of the extensive field of research, see Beate Kortendiek, Birgit Riegraf and Katja Sabisch, eds, *Handbuch Interdisziplinäre Geschlechterforschung* (Wiesbaden: Springer, 2019). For a further context in which, in addition to gender, other distinguishing features that overlap with gender must be taken into account, which is why we speak of intersectionality, see Katrin Meyer, *Theorien der Intersektionalität zur Einführung* (Hamburg: Julius, 2017).

21 An important representative of such a position is Dennett, *Darwin's Dangerous Idea*.

22 Charles Darwin, *The Descent of Man, and Selection in Relation to Sex* (New York: Penguin, 2007), p. 243.

23 In the same place, Darwin cites the following passage from Immanuel Kant: "Duty! Wondrous thought, that workest neither by fond insinuation, flattery, nor by any threat, but merely by holding up thy naked law in the soul, and so extorting for thyself always reverence, if not always obedience; before whom all appetites are dumb, however secretly they rebel; whence thy original?" [In *The Descent of Man*, Darwin quotes from the edition reproduced here, translated by J. W. Semple as *The Metaphysics of Ethics* (Edinburgh, 1836), p. 136. Trans.]

24 Darwin, *The Descent of Man*, p. 253.

25 Ibid., p. 281.

26 Dirk Brockmann, *Im Wald vor lauter Bäumen: Unsere komplexe Welt besser verstehen* (Munich: Deutscher Taschenbuch, 2021), p. 190.

27 Ibid.

28 Here is the study (https://theshiftproject.org/wp-content/uploads/2019/07/2019-02.pdf), and here is a slightly critical discussion (www.br.de/wissen/netflix-klimaschaedlich-planetb-streamen-oekologischer-fussabdruck-berechnung-100.html). Whatever the energy details may be (I am not an expert on this), it is clear that digitization is making a remarkable contribution to global warming that should not be underestimated.

29 See the speculative approach in Bruno Latour, *Facing Gaia: Eight Lectures on the New Climatic Regime* (Cambridge: Polity, 2017), especially the fourth lecture. It would go too far here to trace the many differences between the considerations developed in the main text and Latour's attempt to leave behind both the concept of human being and that of nature. However, I consider the following passage to be the expression of an erroneous conclusion: "[Since] nature is only one element in a complex consisting of at least three terms, the second serving as its counterpart, culture, and the third being the one that distributes features between the first two. In this sense, nature does not exist (as a domain); it exists only as one half of a pair pertaining to one single concept" (Ibid., p. 43). This once again shows Latour's constructivist heritage: he denies the existence of a nature that is independent of us because he thinks that the concept of nature applies to something only in contrast to the concept of culture. This is the same kind of erroneous conclusion as that we cannot imagine a forest independent of us without imagining it, and to deduce from this that there are no forests independent of our imaginings. For we do imagine a forest independent of us, which is no great feat. Latour would have to do more to show that nature as a whole depends in a profound sense on its contrast with culture.

30 See, from the now extensive discussion, Stefano Mancuso, *Die Pflanzen und ihre Rechte: Eine Charta zur Erhaltung unserer Natur* (Stuttgart: Klett Cotta, 2021); and Stefano Mancuso and Alessandra Viola, *Die Intelligenz der Pflanzen* (Munich: Kunstmann Antje, 2015); as well as the universally popular and readable books by Peter Wohlleben (again: *nomen est omen!*), e.g. *Das geheime Leben der Bäume: Was sie fühlen, wie sie kommunizieren – die Entdeckung einer verborgenen Welt* (Munich: Ludwig, 2015). A further in-depth look at the ontology of living things can be found in Merlin Sheldrake, *Verwobenes Leben: Wie Pilze unsere Welt formen und unsere Zukunft beeinflussen* (Berlin: Ullstein, 2020). It is possible that microorganisms such as bacteria could have the greatest influence on shaping the atmosphere according to James Lovelock's

famous (but not well-named) "Gaia hypothesis"; see, for example, *Gaia: A New Look at Life on Earth* (Oxford: Oxford University Press, 1979) and, more disturbingly, *The Revenge of Gaia: Why the Earth is Fighting Back – and How We Can Still Save Humanity* (London: Allen Press, 2006). Latour goes so far as to consider Lovelock to be the Galileo Galilei of postmodernism, who heralded a new age and climate regime. See Latour, *Facing Gaia*. It's beyond the scope of this book to assess critically these positions in any detail. I list this literature here for the reader's guidance.

31 Gilles Deleuze and Félix Guattari, *A Thousand Plateaus: Capitalism and Schizophrenia* (Minneapolis: University of Minnesota Press, 1987).

32 See, e.g., the famous passage in his dialogue *Sophist*: "What the Zeus! Shall we be that easily persuaded that motion and life and soul and thought are truly not present in utterly complete Being? That it neither lives nor thinks; but awful and holy, not possessed of mind, it stands there, not to be moved?" Plato, *Sophist: the Professor of Wisdom* (Indianapolis: Focus, 1996), p. 56. The "interweaving of ideas" (the internet) can be found ibid., p. 71 (translated as "interweaving of the forms").

33 In this way, our hierarchically organized society is always reflected in our image of nature and animals, which is why it was already discussed during Darwin's lifetime whether the British theory of evolution was not essentially a transfer of the socio-economic conditions of modern industrial nations to the animal kingdom. See Friedrich Engels's discussion, in the so-called *Anti-Dühring*, of this question, which he himself answers in a negative way, thus adopting Darwin's doctrine of the "struggle for existence" in a positive way.

34 For a critical discussion of the foundations of evolutionary psychology, see George Ellis and Mark Solms, *Beyond Evolutionary Psychology: How and Why Neuropsychological Modules Arise* (Cambridge: Cambridge University Press, 2017).

35 See the now classic text by Lynn White, Jr., "The Historical Roots of Our Ecological Crisis," *Science* 155/3767 (1967), pp. 1203–7. White also admits that it is not Christianity as such that is responsible for the destruction of the environment (she even ends with praise for St Francis of Assisi, which the current pope should be pleased about) but an agricultural revolution in which a tradition of Christianity is combined with the scientific-technological exploitation of nature. For a different reading of the "religious origin of the ecological crisis" from the spirit of worldly Gnosticism, a late antique religious-philosophical direction to which Christianity also contributed, compare Latour, *Facing Gaia*, with the relevant evidence. I do not consider

this genealogy to be correct either, because an ethic of the destruction of nature does not follow from any particular religious-cosmological worldview. But that is another matter, since the thesis that the ecological crisis (which according to White only really began around 1850) is anyway too far removed historically from Gnosticism and medieval Christianity to lay the blame at their feet (and not, for example, on industrial secular modernity).

36 Max Scheler, *The Human Place in the Cosmos* (Evanston, IL: Northwestern University Press, 2009).

37 See Matthew Cobb, *The Idea of the Brain: The Past and Future of Neuroscience* (London: Basic Books, 2020), and Jessica Riskin, *The Restless Clock: A History of the Centuries-Long Argument over What Makes Living Things Tick* (Chicago and London: University of Chicago Press, 2016).

38 See www.global-solutions-initiative.org/recoupling-dashboard/. I thank Dennis Snower for the many discussions we have had on these topics at the New Institute in Hamburg since February 2020.

39 See www.srf.ch/news/abstimmungen-13-februar-2022/nein-zur-primaten -initiative-eine-interessante-philosophische-frage-mehr-nicht.

40 These quotations are all taken from *Taschenlehrbuch Zoologie*, pp. 3ff. On the relationship of the concept of an animal to modern biology and on the question of how consciousness and life are related, see recently Peter Godfrey-Smith, *Metazoa: Animal Life and the Birth of the Mind* (New York: Farrar, Straus & Giroux, 2021). I would like to thank Jocelyn Maclure for pointing me to this book, to which I had access only after writing the first part, so that it could not really be included in the considerations which led to this book. Here I will just mention the decisive difference: Godfrey-Smith explicitly advocates a biological materialism and monism, according to which all phenomena – and thus also the life of the mind at all its stages – belong seamlessly to nature. He tries to close the gap between our mind and our being animals by placing us in the animal kingdom, a maneuver against which fundamental objections are formulated in the main text. Despite these and other fundamental differences, he elaborates how our conception of animals is provincially shaped by the fact that most of us in everyday life have not yet comprehended the zoological shift from a naïve conception of animals to that of metazoa.

41 The now common view that life has essentially something to do with the second law of thermodynamics, i.e. with the subject of entropy, was made familiar by the great physicist Erwin Schrödinger, *What is Life? The Physical Aspect of the Living Cell* (Cambridge: Cambridge University Press, 2012).

42 Olaf Fritsche, *Biologie für Einsteiger: Prinzipien des Lebens verstehen* (Berlin and Heidelberg: Spektrum Akademischer, 2015), p. 3.

43 See, ibid., the list of properties including examples in the case of systems not identified as alive.

44 It is worth playing a round of the game. There are a variety of easily accessible online platforms where you can watch the game of life, such as at https://playgameoflife.com/.

45 See Max Tegmark, *Life 3.0: Being Human in the Age of Artificial Intelligence* (New York: Vintage Books, 2017).

46 See Markus Gabriel, *The Meaning of Thought* (Cambridge: Polity, 2021).

47 See Michael Thompson, *Life and Action: Elementary Structures of Practice and Practical Thought* (Cambridge, MA: Harvard University Press, 2008).

48 Jacques Derrida, *The Animal That Therefore I Am* (New York: Fordham University Press, 2008). On the gap between humans and other animals from a psychological perspective, see Thomas Suddendorf, *The Gap: The Science of What Separates Us from Other Animals* (New York: Basic Books, 2013).

49 I am aware that this is an inversion of the famous thesis of Arnold Gehlen, who called man a "deficient being" in comparison with animals.

50 See Gabriel, *The Meaning of Thought*.

51 Johann Wolfgang von Goethe, *Faust I & II* (Princeton, NJ, and Oxford: Princeton University Press, 1994), pp. 9–10.

52 This distinction plays an important role in the political theory of the Italian philosopher Giorgio Agamben. He introduces it at the very beginning of his much discussed *Homo Sacer: Sovereign Power and Bare Life* (Stanford, CA: Stanford University Press, 1998). Agamben interprets it in such a way that only *bios* designates the social and political form of life of man, while *zoê* is the term from which later arises the idea of a naked life that state structures must steer, control and subjugate in order to tame it. Agamben himself derives this line of thought, which indeed comes into play in modern political philosophy, especially in Thomas Hobbes's *Leviathan*, from the Aristotelian partial definition of the human being as *zôon politikon*. Since the human being as an animal is first of all only one living being among many, politics – and in modernity this means above all the state – distinguishes it from other living beings, thus dehumanizing its being an animal. On the connection between the concept of the animal and political sovereignty, see Jacques Derrida, *The Beast and the Sovereign*, Volume I (Chicago: University of Chicago Press, 2009). However, one has to keep in mind that these interpretations of the Greek conception are from the outset provided with a valuation, which is especially explicit in Agamben's work. For they intend to show that the state primarily subjugates and controls bare life and thus impermissibly restricts it, which overlooks the fact that there is not only a biopolitics that

turns into the totalitarian and dehumanizes human beings in extreme cases but also a modern health and security system that people willingly accept. At this point, the critical insight, which thinkers such as Michel Foucault, Agamben and Derrida have formulated, is then usually overdone and a narrative is constructed according to which the voluntary acceptance of a state biopolitics is in reality also an expression of a silent and almost invisible power strategy, behind which, however, there are no explicit intentions but, rather, anonymous structures. See, for instance, the one-sided, almost apocalyptic descriptions in Byung-Chul Han, *Psychopolitik: Neoliberalismus und die neuen Machttechniken* (Frankfurt am Main: Fischer, 2014), as well as the sometimes bizarre remarks on health policy in the recent pandemic in Giorgio Agamben, *Where Are We Now? The Epidemic as Politics* (Lanham, MD: Rowman & Littlefield, 2021). While it is true that modern medicine can view the human body as a system of bare life (which corresponds to the third-person point of view), it does not follow that this is never desirable. I would certainly rather go to my dentist than to Aristotle's. A critical analysis of biopolitics should not view it in the horizon of apocalyptic conjecture, because this would blind it to the relevant details and make it think that all biopolitics is an explicit or disguised strategy of subjugation, which is simply not true. The pediatrician who vaccinates children against measles and thus saves lives is not subjugating anyone but, conversely, enabling freedom. The details of the pandemic measures are another topic, but what I call the "virological imperative," which says that we should do everything ethically justifiable to contain the devastating consequences of an uncontrolled contagion, applies.

53 Moses Mendelssohn, "*Phaedon*, or On the Immortality of the Soul," in *Selections from His Writings* (New York: Viking Press, 1975).

54 Amphibian theories can also take on particularly bizarre forms. Derek Parfit, for example, believes that "The body below the neck is not an essential part of us." Derek Parfit, "We Are Not Human Beings," in Stephan Blatti and Paul F. Snowdon, eds, *Animalism: New Essays on Persons, Animals, and Identity* (Oxford: Oxford University Press, 2016), pp. 31–49, at p. 40. For Parfit believes that we are identical only with one part of our body, or ultimately even only with one part of our mind, namely with memory, which holds us together.

55 On this objection, see e.g. John McDowell, "Reductionism and the First Person," in J. Dancy, ed., *Reading Parfit* (Oxford: Oxford University Press, 1997), pp. 230–50.

56 See, for instance, the extensive history of the philosophical attitude on the silence of animals in Élisabeth de Fontenay, *Le Silence des bêtes: la philosophie à l'épreuve de l'animalité* (Paris: Fayard, 1999).

57 Derrida, *The Animal That Therefore I Am*, p. 41.

58 Ibid., p. 61.

59 Ibid., p. 30.

60 Ibid.

61 Not to be confused with another meaning of the same word, also common in animal philosophy, which Corine Pelluchon describes thus: "'Animalism' is a designation for the philosophical, social, cultural, and political movement in which people come together to work for the protection of the interests of animals through their way of life and their collective actions" (*Manifest für die Tiere* [Munich: C. H. Beck], 2020, p. 70).

62 Blatti and Snowdon, p. 2.

63 Eric T. Olson, *What Are We? A Study in Personal Ontology* (Oxford: Oxford University Press, 2007), pp. 23f.

64 See the classic by Derek Parfit, *Reasons and Persons* (New York: Oxford University Press, 1984), a book which is full of science-fiction mind experiments, from beaming to brain transplants.

65 Gabriel, *The Meaning of Thought*.

66 John Locke, *Essay on Human Understanding*, chapter XXVII, 11.

67 Parfit, "We Are Not Human Beings," p. 41.

68 On the theme of the person, see Dieter Sturma, *Philosophie der Person: Die Selbstverhältnisse von Subjektivität und Moralität* (Paderborn and elsewhere: Brill, 1997), as well as Michael Quante, *Person* (Berlin and New York: De Gruyter, 2007).

69 Denis Noble, *The Music of Life: Biology beyond the Genome* (New York and Oxford: Oxford University Press, 2008) and *Dance to the Tune of Life: Biological Relativity* (Cambridge: Cambridge University Press, 2017).

70 Complexity is therefore at least epistemic (from Greek *epistêmê* = knowledge), but maybe also **ontic** (from Greek *on* = being), which is indistinguishable for us in most cases, because we have to understand every complex system as such, which complicates the factual situation in each case.

71 René Descartes, *Discourse on Method and Meditations* (New York: Dover, 2003), p. 38.

72 Ibid. In passing, it follows from this passage and from many other passages and considerations in his work that Descartes is not a modern dualist. He does not mean that a neural state correlates with every mental state, so that body and soul are closely related and yet different. If he were a dualist, he would at most claim that an immaterial substance (the famous *res cogitans*) interacts with our body in a certain way, for example by acting on the pineal gland – a view that is usually attributed to him. But even this metaphysical dualism

cannot be found in Descartes. For him, the activities of thinking, *cogitare*, include not only thinking (*intelligere*) but also feeling (*sentire*), imagination (*imaginari*) and volition (*velle*). However, all of this only exists in the interplay between mind and body, so that Descartes does not reduce the *res cogitans* to an immaterial dimension. I owe this historical insight, which is philosophically significant for our understanding of the modern division between subject and object, to Jens Rometsch, *Freiheit zur Wahrheit: Grundlagen der Erkenntnis am Beispiel von Descartes und Locke* (Frankfurt am Main: Vittorio Klostermann, 2018).

73 Julien Offray de La Mettrie, *Machine Man and Other Writings* (Cambridge: Cambridge University Press, 2003).

74 Hans Driesch, *Die Maschine und der Organismus* (Leipzig: J. A. Barth, 1935), especially pp. 69–72.

75 Ibid., p. 71.

76 Ibid.

77 Ibid., p. 70.

78 Jane Bennett, *Vibrant Matter: A Political Ecology of Things* (Durham, NC: Duke University Press, 2010).

79 Bergson was also involved in an interesting discussion with Einstein. He argued against Einstein that our experience of time cannot be reduced to physical time, about which the physicist presented groundbreaking new insights in the framework of the theory of relativity. See the worthwhile account of the historical context in Jimena Canales, *The Physicist and the Philosopher: Einstein, Bergson, and the Debate That Changed Our Understanding of Time* (Princeton, NJ: Princeton University Press, 2016).

80 Jan Voosholz and Markus Gabriel, eds, *Top-Down Causation and Emergence* (Cham: Springer, 2021).

81 On this topic, see Joachim Bauer, *Das empathische Gen: Humanität, das Gute, und die Bestimmung des Menschen* (Freiburg: Herder, 2021).

82 Christine M. Korsgaard, *Fellow Creatures: Our Obligations to the Other Animals* (Oxford: Oxford University Press, 2018).

83 Corine Pelluchon, *Das Zeitalter des Lebendigen: Eine neue Philosophie der Aufklärung* (Darmstadt: wbg Academic, 2021), p. 80.

84 Korsgaard, *Fellow Creatures*, p. 10.

85 Ibid.

86 Ibid., p. 27.

87 Ibid., p. 20.

88 Ibid., p. 52.

89 Ibid., p. 51.

90 Ibid.

91 Ibid., p. 52.

92 Ibid., pp. 29–33.

93 Korsgaard draws on the approach of her Harvard colleague Thomas M. Scanlon, who in fact makes a similar argument in *What We Owe to Each Other* (Cambridge, MA, and London: Harvard University Press, 1998). But Scanlon, like many others, has taken a realist turn to avoid the sometimes absurd consequences of taking all moral truths, i.e. all answers to ethical questions, as expressions of the rules of construction for a human consensus. See his *Being Realistic about Reasons* (New York: Oxford University Press, 2014).

94 Bruno Latour, *Politics of Nature: How to Bring the Sciences into Democracy* (Cambridge, MA: Harvard University Press, 2004).

95 De Fontenay, *Le Silence des bêtes*.

96 Alice Crary, *Inside Ethics: On the Demands of Moral Thought* (Cambridge, MA, and London: Harvard University Press, 2016), p. 2.

97 Ibid., p. 11.

98 Ibid., pp. 88f.

99 Jakob Johann von Uexküll, *Umwelt und Innenwelt der Tiere* (Berlin and Heidelberg: Spinger Spektrum, 2014), p. 19.

100 See Gabriel, *Moral Progress in Dark Times* and, similarly from an economic perspective, Katharina Lima de Miranda and Dennis J. Snower, "Recoupling Economic and Social Prosperity," *Global Perspectives* 1/1 (2020), 11867.

101 Lorraine Daston and Peter Galison, *Objectivity* (New York: Zone Books, 2010).

102 See Hans Blumenberg, *Wirklichkeiten, in denen wir leben* (Stuttgart: Reclam, 1981).

103 See Markus Gabriel, ed., *Der Neue Realismus* (Berlin: Suhrkamp, 2019), pp. 8–16.

104 Crary, *Inside Ethics*, p. 200.

105 Pelluchon, *Das Zeitalter des Lebendigen*, pp. 19f.

106 See in particular Gabriel, *Moral Progress in Dark Times*; Markus Gabriel and Takahiro Nakajima, *Towards a New Enlightenment* (Shueisha Shinsho, 2020); and Markus Gabriel and Gert Scobel, *Zwischen Gut und Böse: Philosophie der radikalen Mitte* (Hamburg: Edition Körber, 2021).

107 See of course paradigmatically the famous report of the Club of Rome by Dennis L. Meadows et al., *The Limits to Growth: A Report for the Club of Rome's Project on the Predicament of Mankind* (New York: Universe Books, 1972).

108 See Theodor W. Adorno and Max Horkheimer, *Dialectic of Enlightenment* (Stanford, CA: Stanford University Press, 2002).

109 Pelluchon, *Das Zeitalter des Lebendigen*, p. 98.

110 Ibid., p. 99.

111 See www.tagesspiegel.de/politik/angela-merkel-im-wortlaut-wenn-wir-nicht -gerade-aus-stein-sind/14576252.html.

112 However, this is not historically correct. The term "Enlightenment" and what we then know as the Age of Enlightenment emerged in various European contexts in the seventeenth and eighteenth centuries. Enlightenment is much older and appears differently in different places. For example, the Indian economist Amartya Sen rightly points out time and again that religious freedom was introduced on the Indian subcontinent around the time of Akbar in the sixteenth century in order to take account of religious pluralism, which Sen sees as a decisive boost to the Enlightenment. See Amartya Sen: *The Argumentative Indian: Writings on Indian History, Culture and Identity* (London: Picador, 2006). For a differentiated view of the various aspects of the Enlightenment, which is often regarded as "European," see Jonathan I. Israel, *Radical Enlightenment: Philosophy and the Making of Modernity 1650–1750* (Oxford: Oxford University Press, 2002).

113 For Kant's racism, see Immanuel Kant, "On the Different Races of Human Beings," in Kant, *Anthropology, History, and Education* (Cambridge: Cambridge University Press, 2007), pp. 82–97. The question is not whether Kant was a racist (he was) but how his racism fits into his practical philosophy and theory of reason, i.e. whether and how racist his theory of rationality is. But that is another matter.

114 Immanuel Kant, *Lectures on Logic* (Cambridge: Cambridge University Press, 1992), p. 538.

115 David Grene and Richmond Lattimore, eds, *Sophocles I* (Chicago and London: University of Chicago Press, 2013), pp. 320–1.

116 Bruno· Latour, *Facing Gaia: Eight Lectures on the New Climatic Regime* (Cambridge: Polity, 2017).

117 Hans Jonas, *The Imperative of Responsibility: In Search of an Ethics for the Technological Age* (Chicago: University of Chicago Press, 1985).

118 Gabriel, *I am Not a Brain*; *Neo-Existentialism* (Cambridge: Polity, 2018).

119 So also, more recently, Christian List, *Why Free Will is Real* (Cambridge, MA: Harvard University Press, 2019).

120 For those interested in the technical details of the discipline of mereology, which have now been developed logically-mathematically, I can recommend Aaron J. Cotnoir and Achille Varzi, *Mereology* (Oxford: Oxford University Press, 2021). For the context discussed in the main text, see Markus Gabriel and Graham Priest, *Everything and Nothing* (Cambridge: Polity, 2022).

121 Gottlob Frege, "The Thought: A Logical Inquiry," *Mind* 65/259 (1956), pp. 289–311, at p. 291.

122 Eugene Wigner, "The Unreasonable Effectiveness of Mathematics in the Natural Sciences," *Communications on Pure and Applied Mathematics* 13 (1960), pp. 1–14.

123 See also David Deutsch, *The Beginning of Infinity: Explanations That Transform the World* (London: Penguin, 2011).

124 Ludwig Wittgenstein, *Philosophical Investigations*, Part II, §25.

Part II: Social Freedom and the Meaning of Life

1 Arthur Schopenhauer, *The World as Will and Representation*, Vol. II (New York: Dover, 1966), p. 3.

2 On recent estimates of species diversity in the context of his reflections on ecological networks, see Dirk Brockmann, *Im Wald vor lauter Bäumen: Unsere komplexe Welt besser verstehen* (Munich: Deutscher Taschenbuch, 2021).

3 The concept of biopolitics should not be reduced by interpreting all health policy as state intervention in self-determination, as Giorgio Agamben does in his work. This has led him (to put it mildly) to unbalanced historically-diagnostic interpretations of the political measures taken to cope with the pandemic. See Giorgio Agamben, *Where Are We Now? The Epidemic as Politics* (Lanham, MD: Rowman & Littlefield, 2021). Agamben's exaggeration consists in the fact that he considers every state of exception to be funda-mentally unendable, so that he has no theoretical room for the simple option that there are health emergencies to which one can only react biopolitically, as it is precisely this and only this that preserves the social freedom with which Agamben is concerned. Of course, it by no means follows that every measure taken in the fight against Covid was and is proportionate or ethically justifiable.

4 See Markus Gabriel, *I am Not a Brain: Philosophy of Mind for the Twenty-First Century* (Cambridge: Polity, 2017).

5 See Francis Fukuyama, *Liberalism and its Discontents* (London: Farrar, Straus & Giroux, 2022).

6 See, for instance, Charles Taylor, *A Secular Age* (Cambridge, MA: Harvard University Press, 2018), and Markus Gabriel, *Why the World Does Not Exist* (Cambridge: Polity, 2015), chap. 5.

7 See the discussions in Markus Gabriel and Gert Scobel, *Zwischen Gut und Böse: Philosophie der radikalen Mitte* (Hamburg: Edition Körber, 2021).

8 Daniel Kahneman, Olivier Sibony and Cass R. Sunstein, *Noise: A Flaw in Human Judgment* (Boston: Little, Brown, 2022).

9 See the late work by Humberto Maturana and Ximena Dávila Yañéz, *El árbol del vivir* (Santiago: MVP, 2015).

10 See in this connection the speculations of Brian Greene, *Until the End of Time: Mind, Matter, and Our Search for Meaning in an Evolving Universe* (New York: Random House, 2020), chs 3–5.

11 See "Forscher entdecken die Struktur des 'Hoyle-Zustands,'" www.uni-bonn .de/de/universitaet/presse-kommunikation/press-service/archive-press-releases /2012/317-2012.

12 See the report www.scientificamerican.com/article/hoyle-state-primordial -nucleus-behind-elements-life/. A good exposition of the technical reasoning underlying this can be found at http://collaborations.fz-juelich.de/ikp/cgswhp /cgswhp18/program/talks/20.08/Session2/Ulf_Meissner_lifepp4.pdf. In their book *The Anthropic Cosmological Principle* (New York and Oxford: Oxford University Press, 1986), p. 15, the physicists John D. Barrow and Frank J. Tipler define the weak anthropic principle as follows: "The observed values of all physical and cosmological quantities are not equally probable but they take on values restricted by the requirement that there exist sites where carbon-based life can evolve and by the requirement that the Universe be old enough for it to have already done so."

13 Again we should stress that this statement is in no way either speculative or contro-versial. It expresses only the fact that those properties of the Universe we are able to discern are self-selected by the fact that they must be consistent with our own evolution and present existence. WAP [weak anthropic principle] would not neces-sarily restrict the observations of non-carbon-based life but our observations are restricted by our very special nature. (Ibid.)

14 See, for instance, Josef M. Gaßner and Jörn Müller, *Können wir die Welt verstehen? Meilensteine der Physik von Aristoteles zur Stringtheorie* (Frankfurt am Main: Fischer, 2019).

15 See Michael Thompson, *Life and Action: Elementary Structures of Practice and Practical Thought* (Cambridge, MA: Harvard University Press, 2008). For further reading, see Karen Ng, *Hegel's Concept of Life: Self-Consciousness, Freedom, Logic* (New York: Oxford University Press, 2020), and Thomas Khurana, *Das Leben der Freiheit: Form und Wirklichkeit der Autonomie* (Berlin: Suhrkamp, 2017).

16 Aristotle, *Metaphysics* (Indianapolis: Hackett, 2016), p. 206.

17 All of this is probably familiar to some readers from systems theory, which originally comes from the life sciences and was transferred to sociology by Niklas Luhmann, who was particularly prominent in Germany – something with which the famous biologists of the so-called Santiago school (i.e. above

all Humberto Maturana and Francisco Varela), on whose work Luhmann relies, did not agree at all. (Maturana explained the differences to me in detail during a visit to him in Santiago in spring 2018.) The late Maturana went much further than linking life and cognition, as he believed that life begins at a level far below the cell and thus below what the Santiago school had examined as "autopoiesis" on the basis of the cell. See Maturana and Dávila: *El árbol del vivir*.

18 Erwin Schrödinger, *What is Life? The Physical Aspect of the Living Cell* (Cambridge: Cambridge University Press, 2012) (this material was first presented in Dublin in 1943).

19 Robert B. Brandom, *A Spirit of Trust: A Reading of Hegel's Phenomenology* (Cambridge, MA: Harvard University Press, 2019).

20 Hans Jonas, *The Phenomenon of Life: Toward a Philosophical Biology* (Evanston, IL: Northwestern University Press, 2001).

21 See, for an introduction, Gabriel, *Why the World Does Not Exist*, as well as *Sinn und Existenz: Eine realistische Ontologie* (Berlin: Suhrkamp, 2016) and *Fictions* (Cambridge: Polity, 2024).

22 At this point, some may be thinking of the issue of primary (physical) and secondary qualities (which depend on our consciousness and are therefore not yet purely scientifically explorable). This issue played an important role in early modern thought (especially with Galileo Galilei and then in the empiricist epistemology of John Locke and David Hume). This tradition is sometimes understood as if it wanted to reduce all sensual qualities (such as colors) to mathematical, quantitative structures (such as wavelengths). But this is obviously not true, as can be seen from the fact that primary qualities were also referred to as qualities! It would be going too far at this point to pursue the question of whether qualities cannot be described mathematically for some reason, which I myself would currently deny – a topic that I will pursue elsewhere. Briefly, there are good reasons to believe that mathematics can contain qualitative concepts, which in no way implies that physicalism, which traces all fields of sense back to physically explainable processes and structures, is true.

23 I would like to thank Philipp Bohlen for pointing out the research on biofilms. See, for example, Hans-Curt Flemming, Jost Wingender, Ulrich Szewzyk et al., "Biofilms: An Emergent Form of Bacterial Life," *Nature Reviews Microbiology* 14 (2016), pp. 563–75.

24 Robert B. Brandom, *Making it Explicit: Reasoning, Representing, and Discursive Commitment* (Cambridge, MA: Harvard University Press, 1994).

25 See §§ 189–208 of Hegel's *Elements of the Philosophy of Right*, in which he critically examines the emerging economics of his time, especially Adam Smith,

Jean-Baptiste Say and David Ricardo. He reproaches them with a one-sided conception of the human being, which understands the human being as an animal governed by natural necessity, controllable at best economically through the distribution of resources. Already at the beginning of the nineteenth century, Hegel criticizes the idea that the state was ultimately only a system of material resource distribution and thus ultimately only an economic instrument. See the work of social philosopher Lisa Herzog, *Inventing the Market: Smith, Hegel, and Political Theory* (Oxford: Oxford University Press, 2013); and *Reclaiming the System: Moral Responsibility, Divided Labour, and the Role of Organizations in Society* (Oxford: Oxford University Press, 2019).

26 Anyone wishing to explore these topics, which fall into the area of so-called social ontology, should refer to Stephan Zimmermann, *Vorgängige Gemeinsamkeit: Zur Ontologie des Sozialen* (Freiburg im Breisgau: Karl Alber, 2021). Zimmermann puts forward the interesting thesis that a type of action (such as going for a walk or grilling fish) is social even if it takes place without any contact between the actors involved. If two lonely Robinson Crusoes develop a new practice independently of each other, this practice would therefore also be social. For Zimmermann, the social is a question of patterns that occur in reality and in whose existence the mental states of the actors are generally only insignificantly involved. This extreme conclusion does not seem right to me, since the paradigmatic forms of the social arise through implicit and explicit coordination of action, but this already leads into the details of a philosophical argument that I only want to outline here.

27 See in particular Gabriel, *Fictions*, §§ 12–17, on which the considerations in the main text are based.

28 I would like to thank the Boehringer Ingelheim Foundation for organizing a panel on the topic of the vulnerable human being, which took place on 2 November 2021. On that occasion, the renowned immunologist Thomas Boehm introduced some of the details of the individuality of our immune systems, which I pick up on here.

29 In contemporary philosophy, the concept of form of life is mostly linked to Ludwig Wittgenstein's *Philosophical Investigations*. Unfortunately, Wittgenstein does not develop this concept with nearly enough precision to let us see how the "natural history" of humanity he postulates is related to the sociality of the form of life. For a contemporary concept of forms of life anchored in social philosophy (although not, on the other hand, centrally investigating the natural-historical dimension of life, the human being as animal), see Rahel Jaeggi, *Critique of Forms of Life* (Cambridge, MA: Harvard University Press, 2018).

30 Johann Gottlieb Fichte, *Foundations of Natural Right* (Cambridge: Cambridge University Press, 2000), p. 37.

31 Karel Čapek, *The Makropulos Case*, in *Čapek: Four Plays* (London: Methuen Drama, 1999), and Bernard Williams, "The Makropulos Case: Reflections on the Tedium of Immortality," in *Problems of the Self: Philosophical Papers 1956–1972* (Cambridge: Cambridge University Press, 1973), pp. 82–100.

32 Martin Hägglund, *This Life: Secular Faith and Spiritual Freedom* (New York: Pantheon, 2019).

33 See Martin Heidegger, *Being and Time* (San Francisco: Harper, 1962), p. 277.

34 Jean-Paul Sartre, *No Exit, and Three Other Plays* (New York: Vintage, 1955), p. 47.

35 See Susan Wolf, *Meaning in Life and Why it Matters* (Princeton, NJ: Princeton University Press, 2010). For an overview of the anglophone discussion (which is, however, one-sided because it largely ignores the huge global debate outside the English-speaking institutes of philosophy), see the entry "The Meaning of Life" in the *Stanford Encyclopedia of Philosophy*, written by the South African philosopher Thaddeus Metz: https://plato.stanford.edu/entries/life-meaning/.

36 Lisa Herzog, *Freiheit gehört nicht nur den Reichen: Plädoyer für einen zeitgemäßen Liberalismus* (Munich: C. H. Beck, 2014); Ute Frevert: *Kapitalismus, Märkte und Moral* (Munich: Residenz, 2020).

37 Wolf, *Meaning in Life*, pp. 8f.

38 This distinction goes back to John R. Searle, *Making the Social World: The Structure of Human Civilization* (Oxford: Oxford University Press, 2010). For a further development of this distinction with regard to an epistemological concept of reality, see Markus Gabriel, *The Meaning of Thought* (Cambridge: Polity, 2021) and Markus Gabriel and Dominik Krüger, *Was ist die Wirklichkeit? Neuer Realismus und Hermeneutische Theologie* (Tübingen: Mohr Siebeck, 2018).

39 Susan Wolf, "Moral Saints," *Journal of Philosophy* 79/8 (1982), pp. 419–39.

40 Rudolf Carnap, "The Elimination of Metaphysics through Logical Analysis of Language," in A. J. Ayer, ed., *Logical Positivism* (New York: Free Press, 1966), pp. 60–81.

41 Ludwig Wittgenstein, *Tractatus Logico-Philosophicus* (New York: Dover, 1999), p. 107.

42 Ibid., p. 108.

43 Ibid.

44 On these topics, see the classic by Douglas R. Hofstadter, *Gödel, Escher, Bach: An Eternal Golden Braid* (New York: Basic Books, 1979), as well as *I am a Strange Loop* (New York: Basic Books, 2007).

45 See Scarlett Thomas, *The End of Mr Y* (Edinburgh: Canongate, 2006), pp. 17ff.

46 Terry Eagleton, *The Meaning of Life: A Very Short Introduction* (Oxford: Oxford University Press, 2007).

47 Evgeny Morozov, *To Save Everything, Click Here: Technology, Solutionism, and the Urge to Fix Problems that Don't Exist* (New York: Penguin, 2013).

48 Cornelius Castoriadis, *The Imaginary Institution of Society* (Cambridge, MA: MIT Press, 1998), pp. 146–7.

49 Immanuel Kant, *Groundwork of the Metaphysics of Morals* (Cambridge: Cambridge University Press, 1997), p. 24.

50 The existentialist notion of the "draft of such a life plan" is probably found for the first time in Heinrich von Kleist. See his famous letter to Ulrike von Kleist of May 1799 in Philip B. Miller, ed., *An Abyss Deep Enough: Letters of Heinrich von Kleist* (New York: E. P. Dutton, 1982), esp. pp. 26f.

> Such a slavish surrender to the whims of tyrannical fortune is of course most unworthy of a free and thinking person. A free and thinking person does not remain where Chance happens to have pushed him; or if he does, he does so on principle, choosing the better. He feels that one can lift oneself above fate, and even, in a very real sense, master it. He determines by means of Reason what the highest happiness would be for him, lays out a master plan for his life, and strives toward his goal according to securely established principles and with all his powers.

51 This has been shown by the Munich philosopher Axel Hutter in his highly recommended book *Narrative Ontology* (Cambridge: Polity, 2022).

52 Immanuel Kant, *Anthropology, History, and Education* (Cambridge: Cambridge University Press, 2007), p. 111.

53 Georg Wilhelm Friedrich Hegel, *Elements of the Philosophy of Right* (Cambridge: Cambridge University Press, 1991), p. 57.

54 Ibid., p. 59.

55 See in this connection the account in Ute Frevert, *Kapitalismus, Märkte und Moral* (Vienna and Salzburg: Residenz, 2019).

56 See the overview of her approach in Lynn Margulis, *Symbiotic Planet* (New York: Basic Books, 1999).

57 Brockmann, *Im Wald vor lauter Bäumen*, p. 210.

58 Ernst-Wolfgang Böckenförde, *State, Society, Liberty* (New York: St Martin's Press, 1991), p. 45.

59 Immanuel Kant, *Critique of Practical Reason* (Cambridge: Cambridge University Press, 2015), p. 104: "For this reason, again, morals is not properly

the doctrine of how we are to make ourselves happy but of how we are to become worthy of happiness."

60 Ibid., p. 20.

61 Ibid., p. 4.

62 Kant, *Groundwork of the Metaphysics of Morals*, p. 8.

63 Ibid., pp. 8–9.

64 *The Portable Nietzsche*, trans. Walter Kaufmann (New York: Penguin, 1982), pp. 42–3.

65 See Thomas Nagel, *What Does it All Mean? A Very Short Introduction to Philosophy* (Oxford: Oxford University Press, 1987); and *Mortal Questions* (Cambridge: Cambridge University Press, 1991).

66 Thomas Nagel, "What is it Like to be a Bat?," *Philosophical Review* 83/4 (1974), pp. 435–50.

67 Nagel, *What Does it All Mean?*, chap. 10.

68 Nagel, *Mortal Questions*, pp. 11–12.

69 See Karl Jaspers, *Truth and Symbol* (New York: Twayne, 1959) and Nagel, *Mortal Questions*, p. 16: "Those seeking to supply their lives with meaning usually envision a role or function in something larger than themselves. They therefore seek fulfillment in service to society, the state, the revolution, the progress of history, the advance of science, or religion and the glory of God."

70 Nagel, *Mortal Questions*, p. 16.

71 Ibid., p. 21.

72 Gabriel, *I am Not a Brain*; *Fictions*; *Neo-Existentialism* (Cambridge: Polity, 2018).

73 See Markus Gabriel, *Moral Progress in Dark Times* (Cambridge: Polity, 2023), pp. 234–44.

74 On the entanglement of mind and nature in meaning, see the recent and impressive metaphysical and physical outline by Harald Atmanspacher and Dean Rickles, *Dual-Aspect Monism and the Deep Structure of Meaning* (London and New York: Routledge, 2022). I thank Harald Atmanspacher for groundbreaking discussions on the connection between quantum mechanics, field of sense ontology, psychology, and the question of the meaning of life during his visit at the Center for Science and Thought in February 2021 and during my stay at Lake Walen in June 2021.

75 André Laks and Glenn Most, eds, *Early Greek Philosophy*, Vol. II: *Beginnings and Early Ionian Thinkers* (Cambridge and London: Harvard University Press, 2016), pp. 283–5. For those interested, here is the original, which has been translated and interpreted in many ways over the millennia: ἐξ ὧν δὲ ἡ γένεσις ἐστι τοῖς οὖσι, καὶ τὴν φθορὰν εἰς ταῦτα γίνεσθαι κατὰ τὸ χρεών.

διδόναι γὰρ αὐτὰ δίκην καὶ τίσιν ἀλλήλοις τῆς ἀδικίας κατὰ τὴν τοῦ χρόνου τάξιν.

76 Of course, for some time now it has been in vogue to doubt exactly this and to argue for panpsychism, i.e. the assumption that everything real is either psychic or at least proto-psychic. See, for example, Annaka Harris, *Conscious: A Brief Guide to the Fundamental Mystery of the Mind* (New York: Harper, 2019), and Philip Goff: *Galileo's Error: Foundations for a New Science of Consciousness* (New York: Random House, 2019).

77 Markus Gabriel, *Die Erkenntnis der Welt: Eine Einführung in die Erkenntnistheorie* (Freiburg im Breisgau: Karl Alber, 2016), and, for advanced students, both Graham Priest, *Beyond the Limits of Thought* (Oxford and New York: Oxford University Press, 2002), and Markus Gabriel and Graham Priest, *Everything and Nothing* (Cambridge: Polity, 2022).

78 Friedrich Schiller, "Friendship," in Edgar A. Bowring, trans., *The Poems of Schiller* (London: George Bell and Sons, 1875), p. 27.

79 Gabriel, *Moral Progress in Dark Times* (Cambridge: Polity, 2023), pp. 371ff.

80 See Georg Franck, *Ökonomie der Aufmerksamkeit: Ein Entwurf* (Munich: Carl Hanser, 1998), and *Mentaler Kapitalismus: Eine politische Ökonomie des Geistes* (Munich: Carl Hanser, 2005).

Part III: Towards an Ethics of Not-Knowing

1 Max Weber, *The Vocation Lectures* (Indianapolis: Hackett, 2004), pp. 12–13.

2 See the position paper of the New Institute: Markus Gabriel, Christoph Horn, Anna Katsman, Wilhelm Krull, Anna Luisa Lippold, Corine Pelluchon and Ingo Venzke, *Towards a New Enlightenment – The Case for Future-Oriented Humanities*, Discussion Paper no. 2, Hamburg, 2022.

3 See his much discussed major work, Ludwik Fleck, *Genesis and Development of a Scientific Fact* (Chicago: University of Chicago Press, 1981).

4 Wolfram Hogrebe, *Echo des Nichtwissens* (Berlin: Akademie, 2006).

5 This finding, which has been worked out in different fields of science and which is regarded as certain, can be interpreted philosophically in different ways. I myself have argued at length that reality itself is too rich to be grasped in an all-encompassing theory, which I express with the phrase that the world does not exist, which for our purposes means basically that there can be no all-encompassing reality at all (to which we would belong as cognizing living beings grasping it). See Markus Gabriel, *Why the World Does Not Exist* (Cambridge: Polity, 2015), and, on the state of discussion of some of the technical details, see again Markus Gabriel and Graham Priest, *Everything and Nothing* (Cambridge: Polity, 2022).

6 For the partly technical details, see the excellent overview article by James Woodward, one of the leading representatives of interventionist views, in the *Stanford Encyclopedia of Philosophy*: https://plato.stanford.edu/entries/causation-mani/.

7 Christoph Möllers, *The Possibility of Norms* (Oxford: Oxford University Press, 2020). On the socio-ontological limits of such a model, see Gabriel, *Fictions* (Cambridge: Polity, 2024), Part III.

8 See in this sense Holm Tetens, *Gott Denken: Ein Versuch über rationale Theologie* (Stuttgart: Reclam, 2015); Mark Johnston, *Saving God: Religion after Idolatry* (Princeton, NJ: Princeton University Press, 2009), as well as his *Surviving Death* (Princeton, NJ: Princeton University Press, 2010).

9 See in this connection Markus Gabriel, "Der blinde Fleck der Komplexität – Die Wissenschaften in der Krise," in Jakob Augstein, ed., *Follow the Science – aber wohin? Wissenschaft, Macht und Demokratie im Zeitalter der Krisen* (Berlin: Ch. Links, 2022).

10 See in particular Denis Noble, *Dance to the Tune of Life: Biological Relativity* (Cambridge: Cambridge University Press, 2017), as well as his "The Logic of Life: The Public Perception of Science and its Threat to the Values of Society," *Science and Public Policy* 20/3 (1993), pp. 187–92, and *The Music of Life: Biology beyond Genes* (New York and Oxford: Oxford University Press, 2008).

11 Strictly speaking, this is not quite true. For our thinking grasps structures such as logical-mathematical regularities, which is sufficient within the theoretical position taken here to understand the grasping of abstract, i.e. non-changeable and non-material-energetic systems by way of sense perception. See Markus Gabriel, *The Meaning of Thought* (Cambridge: Polity, 2021). However, this is irrelevant for the explanations in the main text at this point insofar as the logical-mathematical structures are not clearly part of nature, or at least not part of changeable nature, if they also limit its change through framework conditions and are therefore causally embedded in a sense that has not yet been conclusively clarified. See, on the discussion of this problem area, Markus Gabriel and Jan Voosholz, eds, *Top-Down Causation and Emergence* (Cham: Springer, 2021), and the book by George Ellis, *How Can Physics Underlie the Mind: Top-Down Causation in the Human Context* (Berlin and Heidelberg: Springer, 2016).

12 Jakob Johann von Uexküll, *Umwelt und Innenwelt der Tiere* (Berlin and Heidelberg: Springer Spektrum, 2014), pp. 113f. Uexküll's ecological thinking is based on distinguishing between the idea of a mechanical "adaptation of the animal to its environment" and the concept of "an incredibly fine adaptation to its environment" (Uexküll, "Wie sehen wir die Natur und wie sieht sie sich

selbst?," *Die Naturwissenschaften* 10/12 [1922], pp. 265–71, at p. 268). It should be noted that he argues against the Darwinism of his time on the basis of his concept of the environment [*Umwelt*], which he distinguishes from that of a merely existing surrounding [*Umgebung*]. With reference to ecological thinking, he explicitly rejects Darwinism when he writes: "In contrast, the doctrine of Darwinism, which wants to trace the organs of all living beings, which are chiseled down to the last detail, back to a random collision of particles dancing around each other without meaning, looks like a dilettantish megalomania" (p. 271).

13 See, for example, Josef M. Gaßner and Jörn Müller, *Können wir die Welt verstehen? Meilensteine der Physik von Aristoteles zur Stringtheorie* (Frankfurt am Main: Fischer, 2019). Of course, the story can also be told quite differently, i.e. without believing that modern physics has refuted classical philosophy of nature. See, for instance, the counter-proposal by Ed Dellian, *Die Rehabilitierung des Galileo Galilei oder Kritik der Kantischen Vernunft* (Sankt Augustin: Academia, 2007). See also Dellian's translation of and especially his introduction to Galileo Galilei, *Discorsi: Unterredungen und mathematische Beweisführungen zu zwei neuen Wissensgebieten* (Hamburg: Felix Meiner, 2015).

14 A large proportion of members of the so-called Stanford school of theory of science argue that there can be no unified physical science. See, for example, Nancy Cartwright, *How the Laws of Physics Lie* (New York: Oxford University Press, 1983), *The Dappled World: A Study of the Boundaries of Science* (Cambridge: Cambridge University Press, 2008), and *A Philosopher Looks at Science* (Cambridge: Cambridge University Press, 2022). See also the contributions in Peter L. Galison and David J. Stump, eds, *The Disunity of Science: Boundaries, Contexts, and Power* (Stanford, CA: Stanford University Press, 1996).

15 Jakob Johann von Uexküll, "Wie sehen wir die Natur und wie sieht die Natur sich selber? Schluß," *Naturwissenschaften* 10/14 (1922), pp. 316–22, at p. 321.

16 Heraclitus, *Fragments* (London: Penguin, 2001), p. 34.

17 See, for example, Hegel's remark on this concept in his *Philosophy of Mind* (Oxford: Oxford University Press, 2017), p. 143 (§ 413): "Just as light is the manifestation of itself and its Other, *darkness*, and can reveal itself only revealing that Other, so too the I is revealed to itself only in so far as its Other becomes revealed to it in the shape of something independent of it."

18 Of course, this raises the question how this entanglement is to be understood. Is it some third that mediates between nature and mind, and, if so, is the third then entangled with one of the two points of reference, which generates consequential problems, or not? For a comprehensive outline of a

theory of entanglement on a level with contemporary philosophy of mind and contemporary physics, see Harald Atmanspacher and Dean Rickles, *Dual-Aspect Monism and the Deep Structure of Meaning* (London and New York: Routledge, 2022).

19 Karl Popper, *Conjectures and Refutations: The Growth of Scientific Knowledge* (New York: Routledge, 2002). See Gaßner and Müller, *Können wir die Welt verstehen?*, pp. 18–25.

20 Ibid., p. 19. Falsificationism in this form is considered refuted in the theory of science. See, on an introductory level, Tim Lewens, *The Meaning of Science: An Introduction to the Philosophy of Science* (London: Basic Books, 2016). Against the idea that we are epistemically trapped in models, which unfortunately we can never compare with reality, see pp. 258–61.

21 Helen Beebee and Huw Price, eds, *Making a Difference: Essays on the Philosophy of Causation* (Oxford: Oxford University Press, 2017).

22 See the overview in Bradley Armour-Garb and Frederick Kroon, eds, *Fictionalism in Philosophy* (Oxford: Oxford University Press, 2017).

23 Hans Vaihinger, *The Philosophy of "As if": A System of the Theoretical, Practical and Religious Fictions of Mankind* (New York: Barnes & Noble, 1968). See, in contrast, the realist theory of fictions in Gabriel, *Fictions*.

24 Wilfrid Sellars, *Empiricism and the Philosophy of Mind* (Cambridge, MA: Harvard University Press, 1997).

25 Thereby, we know neither altogether what we do not know about nature in its being-in-itself nor how far our mathematical methods of measuring the universe reach. We can offer a small, somewhat technical hint for those interested: today there is no comprehensive analytical solution theory for all differential equations. This is a research field of mathematics, which has consequences for the natural sciences, in which differential equations play a central role. For this reason alone, nobody takes in the whole logical space of differential equations, and therefore not all of mathematics either (which is impossible anyway for reasons of metamathematical incompleteness theorems), in order that they might make statements about how the mathematical universe as a whole is constituted. Since the mathematical structures which the physical universe realizes are a subset of the mathematical universe (not everything which mathematics explores is of physical significance), we do not know whether the mathematically measured physical universe is already sufficiently described mathematically in order ultimately to describe the physically measurable universe as a whole. For these and other reasons (which include the current or perhaps even in principle impossibility of experimentally testing certain theories such as string theories), physics, as far as we know,

is not even close to being an all-encompassing theory of the universe. This does not diminish its achievements.

26 Willard Van Orman Quine, *Word and Object* (Cambridge, MA: MIT Press, 1960), p. 275.

27 Scott F. Gilbert, Jan Sapp and Alfred I. Tauber: "A Symbiotic View of Life: We Have Never Been Individuals," *Quarterly Review of Biology* 87/4 (2012), pp. 325–41, at p. 326.

28 I thank my Bonn neuroscience colleagues Heinz Beck and Andreas Zimmer for their explanations of the connection between technological and neuroscientific progress.

29 See the very similar approach of the Ghanaian philosopher Martin Odei Ajei in "An African philosophical perspective on barriers to the current discourse on sustainability," presented at a lecture at the New Institute in Hamburg on December 15, 2021.

30 So also recently Jürgen Habermas: "Reflections and Hypotheses on a Further Structural Transformation of the Political Public Sphere," *Theory, Culture & Society* 39/4 (2022), pp. 145–71. Habermas conceives of a consensual backdrop against which "the whole democratic process consists of a tide of dissent that is stirred up over and over again by the citizens' search for rationally acceptable decisions oriented to the truth … This dynamic of *enduring* dissent in the public sphere likewise shapes the competition between parties and the antagonism between government and opposition, as well as differences of opinion among experts."

31 See in this connection for more detail Markus Gabriel, *Moral Progress in Dark Times: Universal Values for the Twenty-First Century* (Cambridge: Polity, 2022).

32 Jane Flint, Vincent R. Racaniello, Glenn F. Rall, Anna Marie Skalka and Lynn W. Enquist, *Principles of Virology* (Washington, DC: John Wiley, 2015), p. 4.

33 Ibid.

34 Hegel, *Phenomenology of Spirit* (Cambridge: Cambridge University Press, 2018), p. 49.

35 Pierre-Simon Laplace, *A Philosophical Essay on Probabilities* (New York: Wiley & Sons, 1902), p. 4.

36 Ibid.

37 Ibid., pp. 4–5.

38 Ibid., p. 4.

39 Dirk Brockmann, *Im Wald vor lauter Bäumen: Unsere komplexe Welt besser verstehen* (Munich: Deutscher Taschenbuch, 2021), p. 24.

40 Ibid., p. 28.

41 The idea of introducing a third, neutral element in order to take account

of the interconnectedness of mind and nature does not lead to a unified theoretical language insofar as this third element is only partially knowable, so that no sufficiently specific (e.g. mathematical) theory can be formulated in which the complexity of the interconnectedness of mind and nature can be represented. See Atmanspacher and Rickles, *Dual-Aspect Monism*. The theory formulated by Atmanspacher and Rickles considers mind and nature (or the psychic and the physical) as aspects of a neutral reality that is neither mind nor nature. However, this raises the as yet unanswered question as to how a theory of that very third which is neither mind nor nature can be articulated without implicitly or explicitly reverting to the starting position and again characterizing the mind–nature relation unilaterally from nature or unilaterally from mind. The most ambitious attempt so far to spell out two series – one starting from nature and one starting from mind – in order to know the manifestation of the neutral third as a point of indifference in these series was presented by Friedrich Wilhelm Joseph Schelling in his philosophy of nature. However, this was formulated before the achievements of modern, non-classical physics and the development of biology into a modern science (in which, by the way, Schelling had an important part), so that it is unclear to what extent his idea of a philosophy of nature could be maintained under contemporary conditions. And even if a theory of a neutral third could be formulated, one corresponding to all relevant scientific standards, this theory would moreover have to show that it does not proceed reductionistically and lead all fields of sense back to an all-encompassing field of sense which cannot exist. However, the mereological state of affairs is again complex at this point: see Gabriel and Priest, *Everything and Nothing*.

42 See the analysis of the anthropological preconditions of the "exploitation of the animal world," in which a variant of the projection thesis developed in the first part is found, in Theodor W. Adorno and Max Horkheimer, *Dialectic of Enlightenment* (Stanford, CA: Stanford University Press, 2002). However, the authors themselves develop their own mythology of the animal and the animal soul by denying the animals reason, even claiming that "Even the strongest animal is infinitely feeble" (p. 205). Admittedly, they are absolutely right in their insight that the relationship to the 'animal' reflects the social conditions of man.

43 Plato, *The Last Days of Socrates* (London: Penguin, 2010), pp, 36–8.

44 Ibid., p. 38.

45 Thomas Sturm, "'Rituale sind wichtig': Hans-Georg Gadamer über Chancen und Grenzen der Philosophie," *Der Spiegel*, February 21, 2000.

46 There is an extensive philosophical and social science literature on the subject of normativity. A weighty contribution has been made by the jurist and legal

philosopher Christoph Möllers in his book *The Possibility of Norms* (Oxford: Oxford University Press, 2020).

47 Aristotle, *Nicomachean Ethics* (Oxford: Oxford University Press, 2009), p. 23.

48 In his book *Vorgängige Gemeinsamkeit: Zur Ontologie des Sozialen* (Freiburg im Breisgau: Karl Alber, 2021), Stephan Zimmermann shows that this is indeed true even for actions that could be considered entirely non-social.

49 Plato, *Republic* (Indianapolis: Hackett, 1992), pp. 178ff.

50 For introductions from two perspectives, see Markus Gabriel, *Die Erkenntnis der Welt: Eine Einführung in die Erkenntnistheorie* (Freiburg im Breisgau: Karl Alber, 2012), as well as Elke Brendel, *Wissen* (Berlin: de Gruyter, 2013).

51 At this point one can go as deep as one likes into the philosophical theory of facts, but this is unnecessary for our purposes. In particular, it would be imperative to discuss whether there are necessarily existing facts, and thus statements which are true under all circumstances, and whether, conversely, there are also necessarily non-existing facts and thus statements which are not true under any circumstances. The ontology of facts, which deals with this and related questions, is a wide-ranging area of theoretical philosophy. An overview of some aspects of the ontology of facts can be found in the *Stanford Encyclopedia of Philosophy*: https://plato.stanford.edu/entries/facts/.

52 Immanuel Kant, *Groundwork of the Metaphysics of Morals* (Cambridge: Cambridge University Press, 1997), p. 5.

53 Ibid., p. 17.

54 We cannot draw clear boundaries of knowledge, which can be shown employing methods of philosophical epistemology and theory of knowledge in philosophy of language, logics and meta-mathematics, as for instance Graham Priest did in detail in *Beyond the Limits of Thought* (Cambridge: Cambridge University Press, 1995). See also Gabriel, *Die Erkenntnis der Welt*. If you are interested in further details, see yet again Gabriel and Priest, *Everything and Nothing*.

55 Philip N. Howard, *Lie Machines: How to Save Democracy from Troll Armies, Deceitful Robots, Junk News Operations, and Political Operatives* (New Haven, CT: Yale University Press, 2020).

56 See, e.g., the authoritative contributions by Quassim Cassam, *Vices of the Mind* (Oxford: Oxford University Press, 2019), and *Conspiracy Theories* (Cambridge: Polity, 2019), as well as the anthology by Quassim Cassam, Ian James Kidd and Heather Battaly, eds, *Vice Epistemology* (London and New York: Routledge, 2021).

57 Emil Du Bois-Reymond, *Popular Science Monthly* 5 (May 1874), p. 28.

58 Ibid., p. 31.

Index